H. W. Roesky · K. Möckel

Chemical Curiosities

© VCH Verlagsgesellschaft mbH, D-69451 Weinheim (Federal Republic of Germany), 1996

Distribution:

VCH, P.O. Box 10 11 61, D-69451 Weinheim (Federal Republic of Germany)

Switzerland: VCH, P.O. Box, CH-4020 Basel (Switzerland)

United Kingdom and Ireland: VCH (UK) Ltd., 8 Wellington Court, Cambridge CB1 1HZ (England)

USA and Canada: VCH Publishers, Inc., 337 7th Avenue, New York, NY 10001 (USA)

Japan: VCH, Eikow Building, 10-9 Hongo 1-chome, Bunkyo-ku, Tokyo 113 (Japan)

ISBN 3-527-29414-7

H. W. Roesky · K. Möckel

Chemical Curiosities

Spectacular Experiments and Inspired Quotes

With a Foreword by Roald Hoffmann

Translated by

Prof. Dr. T. N. Mitchell and
Prof. Dr. W. E. Russey

VCH

Weinheim · New York
Basel · Cambridge · Tokyo

Prof. Dr. Herbert W. Roesky
Institut für Anorganische Chemie
Universität Göttingen
Tammannstraße 4
D-37077 Göttingen
Germany

Prof. Dr. Klaus Möckel
Thälmannstraße 10
D-99974 Mühlhausen
Germany

Translators: Prof. Dr. T. N. Mitchell
Prof. Dr. W. E. Russey

Published jointly by
VCH Verlagsgesellschaft mbH, Weinheim (Federal Republic of Germany)
VCH Publishers, Inc., New York, NY (USA)

Editorial Director: Dr. Anette Eckerle, Dr. Thomas Kellersohn
Production Manager: Dipl.-Ing. (FH) Hans Jörg Maier

Cover Picture: Chemoluminescence of singlet oxygen; courtesy of Prof. Dr. M. Kasha, Florida State University.

Deutsche Bibliothek Cataloguing-in-Publication Data:
Roesky, Herbert W. :
Chemical curiosities : spectacular experiments and inspired
quotes / H. W. Roesky ; K. Möckel. With a foreword by Roald
Hoffmann. Transl. by T. N. Mitchell and W. E. Russey. –
Weinheim ; New York ; Basel ; Cambridge ; Tokyo : VCH, 1996
  Dt. Ausg. u.d.T. : Roesky, Herbert W. : Chemische Kabinettstücke
  ISBN 3-527-29414-7
NE: Möckel, Klaus:

Composition: Fa. Mitterweger Werksatz GmbH, D-68723 Plankstadt. Printing: betz-druck gmbh, D-64291 Darmstadt. Bookbinding: Wilhelm Osswald & Co., D-67433 Neustadt.
Printed in the Federal Republic of Germany.

Foreword

Where shall we position the masterpieces of the chemical demonstrator's art? Somewhere between white magic and science. Somewhere between gripping theater and chemistry. Somewhere between circus and the Zen *koan* that bestirs the dormant knowledge in a student's mind.

We know there is no deeper silence in a classroom than that which accompanies the first seconds of a demonstration. Theater directors and nervous concert hall managers envy us those natural seconds of rapt attention. The auditorium is hushed, awaiting change. The demonstrator does not fail to provide it, with color, flame, smoke, and explosion. There ensues – catharsis for the lecturer, a catering to all the senses of the audience, and sometimes the only thing the students remember from a course.

And yet, yet ... in the hands and words of a good teacher, so many more and valuable things may flow from that demonstration:

... a repeated metaphor for the heart of chemistry, which after all was and is the art, craft, science, and business of substances and their *transformations,*

... the punctuation point, the summary of the essence of a principle,

... the intellectual alarm clock, provoking chemical enlightenment,

... the essential question, whose asking makes a teacher delirious if it arises in the minds of just a few: "What *is* happening?"

... a return to the real world; science works with flights of imagination interspersed with contact with mundane but ever-puzzling reality. The process of teaching naturally stresses symbols – words, concepts, and theories. The demonstration touches the earth. It may be staged, but it is tangible.

The remarkable achievement of Herbert Roesky's and Klaus Möckel's book is the linkage it achieves between the world of the human spirit, expressed in literature and historical continuity, and the art of chemical demonstration. One expects Goethe to move freely in the pages of "Chemical Curiosities", but Whitman, Nietzsche, Thomas Mann, Salvador Dali, Montaigne, and the prophet Jeremiah! They serve too, authentically and ingeniously, in the authors' deeply humanistic approach to science. The chemical and literary strands of this book are so ably intertwined.

One thing is missing in this book. It is the protagonist, you – teacher and actor – daring to try that demonstration. Do it!

Ithaca, May 1994   *Roald Hoffmann*

Preface

Both science and art involve highly demanding creative activity. Both are cognitive processes which resemble one another in their creativity and craftsmanship. They differ perhaps only in intensity, in the degree of workability or in their use of symbolism. Thus the research on the composition and the inner structure of the dead and living matter surrounding us, which has been carried out with such great success by chemists in the last two hundred years, requires a series of steps which are cleverly devised, logical, rationally predictable and controlled; the composer needs exactly the same type of steps in order to produce a symphonic masterwork, as does the painter in transferring his artistic ideas to canvas. Intuitive thought processes and the figurative grasp of a problem are considered to be essential elements of the artistic creative process, but are linked much less often with the gathering of scientific knowledge. Yet the construction of the periodic table of the elements, the determination of the series of reactions occurring when a radioactive uranium nucleus undergoes fission, or the magnificent development of routes for the total synthesis of chlorophyll or vitamin B_{12} require just as high a degree of fantasy and of simultaneous and total comprehension as those presented by Hermann Hesse in the "The glass bead game" or Thomas Mann in the novel "Dr. Faustus". Just as Hegel refers on the artistic plane to the "harmonic unity" of different or even contradictory elements of a work of art, the structures of buckminsterfullerene or the crown ether complexes are the result of magnificent chemical feats which allow art to unfold in science and science in art. Wilhelm Ostwald once said that regularity is the foundation for harmony and beauty, and the book he wrote which was published in 1922 bears the title "Die Welt der Formen. Entwicklung und Ordnung der gesetzlich-schönen Gebilde" (loosely translated as: "The World of Shapes. Development and Order of Legitimately Beautiful Creations"). This phrase expresses precisely the unity of scientific research and artistic expectation which was felt and strived for both by himself and by many other great chemists. As early as the end of the 18th century Goethe, who will quite often be cited in this book, made use of his excursions in natural science (which continued until 1810 or 1815) to bind together again that which was being torn apart by the mighty currents of the time: analytical reasoning and creative fantasy, artificial experiment and real experience, abstract concepts and sensual conceptions. Goethe refers to the artificial as an experiment. Each of us chemists knows how good chemical teaching, reinforced by attractive experiments, has made a lasting impression on us and how our own experimental work has inspired us to imaginative reflection on the nature of our science. Thus chemical discoveries and the understanding of chemical concepts are unthinkable without the use of experiments. It therefore seems apt to quote from Liebig's "Introduction to lectures on experimental chemistry" (1852).

There is no art so difficult as the art of observation: it requires a trained, sober spirit and highly requires disciplined experience, available only through practice; for the observer is not the one who sees of what components a thing consists, and the relationship between the components and the whole. Many overlook half of what is there through carelessness, another reports more than what he sees in that he confuses it with that which he imagines, yet another sees the parts that make up the whole but throws things together that must instead be separated. Once the observer has established the basis of his phenomenon, and is in a position to assemble the preconditions, he then proves the validity of his observations by attempting to summon forth the phenomena according to his own will in the form of an experiment. Conducting a series of experiments often means dissecting a thought into its components and testing the result through a sensory phenomenon. A natural scientist conducts experiments in order to verify the truth of his understanding; he conducts experiments to demonstrate a phenomenon in all its many aspects. If the wishes to establish that a series of phenomena are all based on the same cause, then he seeks a simple expression of the same, which in this case is referred to as a natural law. We speak of a simple characteristic as a natural law when it serves to explain one or more natural phenomena.

Chemical experiments possess a special, unmistakeable appeal for both the experimentalist and his audience, not only in the scientific process but of themselves. In the United States so-called "magic shows" enchant both college students and the public. The experimental lectures which one of us has presented to a wide range of audiences in all parts of the Federal Republic of Germany under the title "Chemische Kabinettstücke" have been enjoyed by experienced chemists from industry, by science teachers, by students and by scientifically interested members of the general public. We are happy that these lectures have received such a favorable response and are thus glad to fulfil the often expressed wish that these "Chemische Kabinettstücke" should be published in book form. It was clear to us that this would be a difficult task because of the standard which we wanted to attain in this project. Every experimental lecture thrives not only on the skill of the experimentalist but also on his ability to improvise and to build bridges "effortlessly" to other themes of our time. We also wanted to combine exact experimental procedures with the background knowledge necessary for the student and at the same time to offer the expert interesting stimuli on the history of chemistry or give him guidance as to where he could find information on additional aspects of the experiment. And finally, in the sense of our opening remarks, we intend to integrate elements of art, in particular aesthetic aspects but also the variety of forms which it can take, with our chemical topics. Poems, aphorisms, anecdotes and descriptions of daily occurrences will be used to make clear the central role of chemistry in our culture and civilisation.

However, we can not always follow Nietzsche when he demands "There where thou standest, dig deep! For the source is below." We shall forego a detailed didactic discussion of the experimental procedures, because the experiment must primarily work as a complete entity. Thus comments on each individual step of the experiment would often do more damage than they could possibly be of help. In some cases a detailed theoretical interpretation of the course of the reaction, as for example in the case of oscillating systems or adsorption effects in chemical separations, would be more likely to confuse the untrained reader, so that in such cases we merely refer to the original literature. In some cases an exact theoretical treatment of the experiment is not even possible. Here we follow the precept of Thomas Aquinus: "amazement is a thirst for knowledge." We have not arranged the "Chemical Curiosities" in a rigid order. Special effects such as the variety of forms in which salt crystals can exist, the multitude of colors formed by acid-base and complexation reactions in solution, the energy evolved or consumed when substances are mixed or the interaction between substance, light and heat will however be treated together. We have made considerable use of B. Z. Shakhashiri's valuable compendium "Chemical Demonstrations, A Handbook for Teachers of Chemistry" (which now encompasses four volumes). We have also evaluated with pleasure the fine experimental contributions to "Journal of Chemical Education" and "Chemie in unserer Zeit". Our particular thanks go to Henry Fraatz of the Institut für Anorganische Chemie in Göttingen for his painstaking help and his experimental skill in the design and preparation of the experiments. He also made valuable contributions to the development of some new experiments. We also thank all those who were involved in preparing the manuscript, the diagrams and the photographs. We are grateful to the VCH Verlagsgesellschaft for constructive discussions and for their readiness to comply with our requests. We offer our book to the interested reader in the hope that he or she, together with all those who value the science of chemistry, may enjoy the beauty of the experiment and treasure its value in the eternal struggle between the use of the treasures provided by nature and the admiration of their beauty.

Herbert W. Roesky
Klaus Möckel

Göttingen, May 1994

Contents

1 Pictures that Paint Themselves 1

2 A Golden Rain of Lead Iodide Crystals 5

3 Fractal Structures Made of Silver 7

4 A Lead Tree 10

5 A Silver Mirror 13

6 A Copper Mirror 15

7 A Mercury Mirror 18

8 Pharaoh's Snake 20

9 A Chemical Garden 23

10 Self-Igniting Iron 25

11 The Passivation of Iron 27

12 Fireworks Ignited by Ice 29

13 The Aluminum-Iodine Reaction 31

14 Lightning under Water 33

15 Growling Gummy Bear 35

16 Self-Igniting Wood Wool 37

17 Fire without Matches 39

18 White Phosphorus, a Very Dangerous Chemical 41

19 Fireworks for a Garden Party – Red and Green Fire 45

20 Red, Green, and Yellow Fire without Emission of Sulfur Dioxide . . 48

21 Artificial Fog 50

22 Mushroom Clouds 52

23 The Thermite Process 54

24 Antimony Triiodide 60

25 The Many Colors of Vanadium 62

26 The Stepwise Reduction of Potassium Permanganate in an
Alkaline Medium 64

27 A Reversible Blue-and-Gold Reaction 66

28 The Two-Color Formaldehyde Clock 68

29 The Colors Black, Red, and Gold 70

30 Reactions with Iodine 73

31 Methylene Blue: A Dyestuff that Made Medical History 76

32 The Volcano Experiment 78

33 Catalytic Decomposition of Ammonia in the Presence of Oxygen . . 80

34 Hydrogen Peroxide and Blood. 82

35 Decomposition of Hydrogen Peroxide in the Presence of
Manganese Dioxide 86

36 Decomposition of Hydrogen Peroxide by
Potassium Permanganate. 88

37 Decolorization of Permanganate Solution by Hydrogen Peroxide. . . 90

38 The Bleaching of Hair 92

39 Invisible Inks. 95

40 A Magic Box 97

41 A Weather Station. 100

42 Crown Ether Inclusion Compounds 102

43 Color Effects due to Ligand Exchange in Nickel Complexes. 105

44 A Simple Separation of Cobalt and Nickel Salts 108

45 The Reaction of Iron(III) Chloride with Hydroxybenzenes 110

46 Five Colors from One Solution .112

47 Color Effects in Aqueous Systems Containing Divalent Metal Ions
Derived from Selected 3d Elements. 114

48 Color Reactions as a Test for Solvents 117

49 Equilibrium Reactions of Copper and Cobalt Complexes. 120

50 The Colors of the Rainbow. 123

51 Plant Dyes as Universal Indicators 126

52 Chemical Equilibria in Mineral Water 129

53 Dry Ice and Indicators. 131

54 Self-Organization in Solution 133

55 Acidic and Basic Salts 135

56 The Amphoteric Behavior of Aluminum 137

57 The Ammonia Fountain 139

58 Stars and Stripes Forever 142

59 Ion-Exchange Resins 144

60 A Three-Layer Liquid 176

61 Some Very Different Cocktails. 178

62 The Nernst Distribution Law 180

63 Separation of Leaf Pigments by Column Chromatography 183

64 Separation of the Colored Inks from Felt-Tip Pens 186

65 Chemoluminescence 188

66 Two-Color Chemoluminescence. 191

67 Chemoluminescence with Oxalyl Chloride 194

68 Singlet Oxygen 196

69 Generation of Singlet Oxygen in the Presence of Dyestuffs 199

70 The Mitscherlich Test. 201

71 The Chemoluminescence of Phosphorus 203

72 Chemoluminescence with Oxalic Esters. 205

73 Hemoglobin Chemoluminescence. 208

74 Developing of a Picture with Light 210

75 Where There Is Light, There Is Also Shadow; Experiments with
   UV Light 212

76 A Blueprint. 214

77 Photochemical Reduction of a Thiazine Dye 216

78 Whiter Than White 218

79 A Simple Luminophore 219

80 The Setting Sun 221

81 Mercury Beating Heart 223

82 Gallium Beating Heart 225

83 How to Make Batteries from Fruit and Vegetables 228

84 Colors around the Cathode 230

85 How to Turn Aluminum into Hoarfrost 232

86 How to Clean Silver Cutlery 234

87 Experiments with Liquid Nitrogen 236

88 Cigars Burn Better in *Liquid* Air! 239

89 Demonstration of the Meissner-Ochsenfeld Effect:
    A Hovering Superconductor 240

90 A Highly Exothermic Reaction 243

91 Highly Endothermic Reactions 245

92 An Eruption Caused by Zinc and Sulfur 247

93 Thermochromism 249

94 A Simple Experiment to Demonstrate the "Greenhouse Effect" 251

95 A Barking Dog 254

96 A Bromate-Malonic Acid Oscillation Process Catalyzed by
    Mn^{2+} Ions 256

97 A Green-Blue-Red Belousov-Zhabotinsky Reaction 258

98 An Oscillating Platinum Wire 260

99 Green-Red-Yellow: An Unusual Traffic Light 262

100 A Flashing Blue Light 264

101 The Döbereiner Cigarette Lighter: Physical and
     Chemical Properties of Hydrogen 266

102 The Landolt Experiment 270

103 "Home-Brewed Beer" 275

104 "Artificial Coke" 277

105 The Chlorine-Hydrogen Reaction 279

106 Big Bang in a Tin Can 282

107 Gas Explosions 284

108 The Reaction of a Mixture of Acetylene and Air 286

109 Oxyhydrogen Gas in Soap Bubbles. 289

110 Nitrogen Triiodide 292

111 Minting Coins using Potassium Chlorate 294

112 Combustion with Emission of Sparks 296

113 The Charcoal Dance: Reactions of Charcoal and Sulfur with
    Fused Potassium Nitrate. 298

114 Dancing Fire 300

115 A Burning Gel 303

116 Borates. 304

117 Ethyl Acetate 306

118 Esters as Natural Perfumes 308

119 Reactive Aldehydes 311

120 How to Dissolve Polystyrene Foam 315

121 Hoarfrost in a Glass 316

122 Sulfur Crystals. 318

123 Giant Crystals 321

124 A Handkerchief Flambé, and How (Not) to Burn a Banknote. 324

    Cited Personalities 329

    Subject Index. 333

1

Pictures that Paint Themselves

Friedrich Ferdinand Runge (1794–1867) writes as follows in his book "*Zur Farben-Chemie, Musterbilder für Freunde des Schönen und zum Gebrauch für Zeichner, Maler, Verzierer und Zeugdrucker*":

In the case of the chemical investigations known as decompositions or disintegrations, it is important first to establish what substances one is working with or, speaking chemically, what substances are present in a particular composite or mixture. For this purpose one takes advantage of so-called counteracting agents; i.e., substances with specific properties or peculiarities, and which one knows in detail on the basis of reports or personal experience, so that the changes they induce or suffer constitute the language they speak, thereby informing the researcher that this or that particular substance is present in the mixture in question.

In this way every little picture is associated with a history of its origins, experienced according to chemical law, and one could write about each a small treatise, discussing therein the peculiar chemical behavior of each substance and achieving much clearer distinctions than is possible with the usual mixing and dissecting methods in chemical lectures. Chemical facts already discussed, and those yet to be discussed, which so quickly disappear in a so-called chemical experiment, lie spellbound here before us, the equivalent of a true-to-life map of some specific region of chemistry.

Runge dedicated his book to His Majesty *Friedrich Wilhelm IV*, who wrote the following letter of thanks:

I herewith express to you my great appreciation for the submitted first part of your Chemical Samples, the joyous receipt of which has already been assured to you by the acclamation that a first glance at it evoked from me.

Bellevue, the 30th of October, 1850   *Friedrich Wilhelm*

Title page of the book by *Friedrich Ferdinand Runge*, „*Zur Farben-Chemie, Musterbilder für Freunde des Schönen und zum Gebrauch für Zeichner, Maler, Verzierer und Zeugdrucker*".

Apparatus

Stand, 3 bosses, 2 clamps, aluminum (or iron) rod about 30 cm in length, wooden board ca. 30 × 40 cm, 2 pieces of balsawood ca. 12 × 2 × 2 cm, 4 drawing pins, special paper 14 × 14 cm (chromatographic paper). Prior to use this paper is impregnated by immersing it in a 3 % aqueous solution of copper sulfate, $CuSO_4 \cdot 5 H_2O$. Eight dropping flasks, Pasteur pipettes, glass tube ca. 6 cm in length as a holder for the Pasteur pipette, safety glasses, protective gloves.

Chemicals

4 dropping flasks containing:
a) first developer solution (10 % aqueous solution of ammonium dihydrogen phosphate, $NH_4H_2PO_4$)
b) second developer solution (3 % aqueous solution of potassium hexacyanoferrate, $K_4[Fe(CN)_6]$)
c) tap water
d) salt solution (made by dissolving a pinch of salt in 10 mL tap water)
   A further 4 dropping flasks are filled with blue, red, yellow and green food colorings.

Experimental Setup

The two pieces of balsawood are attached to the wooden board, about 15 cm apart. The paper is now attached at its edges to the two pieces of balsawood. The paper should lie flat and horizontally (the slight tilt of the board during the demonstration has virtually no effect on the "development" of the picture). The board is then attached at an inclination of about 45° to the stand by means of a clamp; an aluminum rod is also attached to the stand. This aluminum rod carries a clamp which holds the glass tube used to maintain the Pasteur pipette vertically above the mid-point of the special paper. The solution to be applied is slowly drawn out of the pipette onto the chromatographic paper. This prevents the formation of a "blob", which would result if the paper were kept in the horizontal during a dropwise application (colored figure 1).

Procedure

The special paper used in this experiment is always treated initially with the first developer solution; this is applied using a Pasteur pipette. The colors are now applied: interesting effects can be obtained when a pause is allowed to elapse between the application of the various colors used. These effects can be made even more interesting by hindering the capillary flow, for example by cutting patterns out of the special paper with a knife or by applying a few drops of salad oil to the paper.

"Development" of two pictures in a relatively short time (colored figure 2):

eight-minute run	eleven-minute run
1. First developer solution	1. First developer solution
2. **Blue** coloring	2. **Red** coloring
3. Salt solution (apply twice)	3. Water (apply 3 times)
4. Second developer solution	4. **Green** coloring
5. Salt solution	5. Water
6. **Yellow** coloring	6. Salt solution (apply twice)
7. Water (apply 3 times)	7. Second developer solution
8. Salt solution (apply 3 times)	8. **Yellow** coloring

Explanation

The dyes in the colorings are separated according to the rate at which they migrate during the chromatographic procedure.

Wider zones are normally formed when salt solutions are used for development, as the displacement of the colors towards the edge of the picture is weaker. In contrast, water washed the color strongly from the center of the picture towards the edges.

Waste Disposal

Liquid residues can be flused down the drain, while the pictures should be treated like normal household waste.

Reference

1 G. Harsch, H.H. Bussemas, *Bilder, die sich selber malen*, DuMont, Cologne, **1985**.

2

A Golden Rain of Lead Iodide Crystals

The first who likened painting and poetry to each other must have been a man of delicate perception, who found that both arts affected him in a similar manner. Both, he realised, present to us appearance as reality, absent things as present; both deceive, and the deceit of either is pleasing.

A second sought to penetrate to the essence of the pleasure, and discovered that in both it flows from one source. Beauty, the conception of which we at first derive from bodily objects, has general rules which can be applied to various things: to actions, to thoughts, as well as to forms.

A third, who reflected on the value and the application of these general rules, observed that some of them were predominant rather in painting, others rather in poetry; that, therefore, in the latter poetry could help out painting, in the former painting help out poetry, with illustrations and examples.

The first was the amateur; the second the philosopher; the third the critic.

Gotthold Ephraim Lessing, "Laokoon", Preface

Roald Hoffmann reminds us of the following metaphor in his poem *"Men and Molecules"*:[1]

Men (and women) are not as different from molecules as they think ...

Safety Measures

It is very important to prevent lead salts from coming in contact with the skin!

Apparatus

One-liter round-bottomed flask, two small beakers, lamp with 100 W bulb (as a spotlight), black cardboard, water bath, stand, safety glasses, protective gloves.

Chemicals

Potassium iodide, lead acetate, concentrated acetic acid.

Experimental Procedure

An aqueous solution of lead acetate is mixed with potassium iodide. The resulting suspension is allowed to stand for a short time and the solution decanted from the yellow precipitate. This is added (without drying it) to the round-bottomed flask which has been filled with ca. 800 mL of hot water (70–80 °C). A little concentrated acetic acid is added to make the solution acidic. A sufficient amount of lead iodide is taken to produce a homogeneous solution, which is almost colorless at this temperature. (A saturated solution has a concentration of about 9 mmol/L, so that slightly under 3 g of lead iodide are required). The flask is now swirled slowly and cooled under a stream of cold water from the tap. After a few seconds beautiful shiny golden crystals begin to form; these slowly fill the flask. If the lamp is allowed to shine on the flask against a black background (such as the cardboard) the crystallisation process appears to the viewer as "golden rain" in a blaze of beautiful color. The optical effect is even better if the room is darkened (colored figure 3).

Explanation

A yellow precipitate of amorphous lead iodide is formed immediately at room temperature:

$$(CH_3COO)_2Pb + 2\ KI \rightarrow PbI_2 + 2\ CH_3COOK$$

The solubility of lead iodide at this temperature is about 1 mmol/L. At the temperature of the experiment the precipitate dissolves and crystallises out again on cooling.

Waste Disposal

Because of the extremely low solubility product of lead iodide at room temperature ($K_{sp} \approx 3 \times 10^{-9}$) it is only necessary to filter off the contents of the flask, to wash the precipitate several times with water and to transfer it to a break-proof vessel used for collecting less toxic inorganic waste. The waste water can be flushed down the drain.

Reference

1 R. Hoffmann, *The Metamict State*, University of Central Florida Press, Orlando, **1987**, 43.

3

Fractal Structures Made of Silver

But one thing I will say: such weirdnesses are exclusively Nature's own affair, and particularly of nature arrogantly tempted by man.

Thomas Mann, "Doctor Faustus"

The term fractal, introduced by *B. B. Mandelbrot* almost 20 years ago, is of central importance for understanding the basic growth and self-organisation processes in nature. *H.-O. Peitgen* et al.[1] cite *Mandelbrot* as follows:

Why is geometry often described as cold and dry? One reason lies in its inability to describe the shape of a cloud, a mountain, a coastline, or a tree. Clouds are not spheres, mountains are not cones, coastlines are not circles, and bark is not smooth, nor does lightning travel in a straight line... Nature exhibits not simply a higher degree but an altogether different level of complexity. The number of distinct scales of length of patterns is for all purposes infinite.

The existence of these patterns challenges us to study those forms that Euclid leaves aside as being formless, to investigate the morphology of the amorphous. Mathematicians have disdained this challenge, however, and have increasingly chosen to flee from nature by devising theories unrelated to anything we can see or feel.

Dendritic fractals can very easily be generated via electrolytic reduction of silver ions.

Apparatus

Overhead projector, Petri dish 15 cm in diameter and 2 cm high, stand, bosses, clamps, alligator clips, electric cable, DC transformer (22 V), paper clips (or iron wire) as electrodes, filter paper (diameter 12.5 cm), safety glasses, protective gloves.

Chemicals

0.1 mol/L $AgNO_3$ solution, aqueous ammonia, dilute hydrochloric acid.

Silver Fractals

Experimental Procedure

The Petri dish is placed on the overhead projector. About 200 mL of a 0.1 mol/L solution of silver nitrate in aqueous ammonia are poured in (8.5 g $AgNO_3$ are dissolved in 400 mL H_2O, the solution is made alkaline by adding aqueous ammonia, and water is added to give a total volume of 500 mL). The positive electrode (anode), a partially opened up paper clip, is attached to the edge of the dish in such a way that it is immersed in the solution; the negative electrode (cathode), in this case a completely straightened out paper clip, is placed in the middle of the dish. The metal point is lowered until it just touches the surface of the solution (surface tension); if it is immersed too far no fractals will be formed. When the 22 V DC current is switched on, a dendritic fractal structure consisting of metallic silver is formed immediately at the cathode. It grows towards the anode at the phase boundary between the air and the solution. The structure can be removed carefully using a round piece of filter paper and the experiment repeated.[2, 3]

Explanation

The solution of silver nitrate in aqueous ammonia is reduced to metallic silver during the electrolysis.

$$Ag^+ + e^- \rightarrow Ag$$

Waste Disposal

Metallic silver can be reused. **Caution!!** Solutions of silver salts in aqueous ammonia cannot be kept for long periods of time, since explosive Ag_3N can be formed. The residues should therefore be kept slightly acidic. The mixture is treated in a beaker with 5 mol/L hydrochloric acid and reduced to metallic silver by means of zinc rods.

References

1 H.-O. Peitgen, P. H. Richter, *The Beauty of Fractals,* Springer Verlag, Berlin, Heidelberg, New York, Tokyo, **1986**.
2 W. V. Ligon, Jr., *J. Chem. Educ.,* **1987**, *64*, 1053.
3 L. P. Silverman, *J. Chem. Educ.,* **1992**, *69*, 928.

4

A Lead Tree

In the 18th century the so-called *Arbores Dianae* were well-known symbolic figures; their introduction was due to the rediscovery of Greek mythology. Thus Diana, Apollo's sister, was related to the moon and thus to the element silver. We have seen in the previous experiment that it is easily possible to generate silver trees, which are formed as fractal structures during the electrolysis of silver nitrate in aqueous ammonia; impressive treelike structures can also be generated using other metals. The following is the procedure described for the lead tree *Arbor Saturni* by the Thuringian apothecary and chemist *J. C. Wiegleb* (1732–1800) from Langensalza[1]:

> *One first dissolves an ounce of sugar of lead (lead acetate) in twelve ounces of distilled water and then filters the liquid to such a clarity that it looks as light as pure spring water. Then one melts approximately one-quarter or one-half pound of zinc in a crucible or in a strong iron spoon, pouring minuscule portions of this into cold water in a wooden vessel. It often happens in the process that the zinc creates tree-like figures. One then seeks out the best of the pieces, that which most resembles a branch, binds it with a delicate linen thread, and hangs it precisely in the center of a sugar glass, with the twigs extending upward, such that it does not touch the bottom. Then the previously described pale solution is poured into the same glass so that the tiny metal branch comes to be suspended in the middle, and the glass is placed in a quiet spot .... The tiny tree (becomes) metallically covered with the most beautiful, sparkling lead crystals and extraordinarily lovely to behold ....*

Safety Precautions

Great care must be taken to prevent the (soluble) lead salts from coming in contact with the skin!

Apparatus

Stand with double boss, overhead projector, 200-mL Erlenmeyer flask, 250-mL measuring cylinder, balance, spatula, dropping pipette, safety glasses, protective gloves.

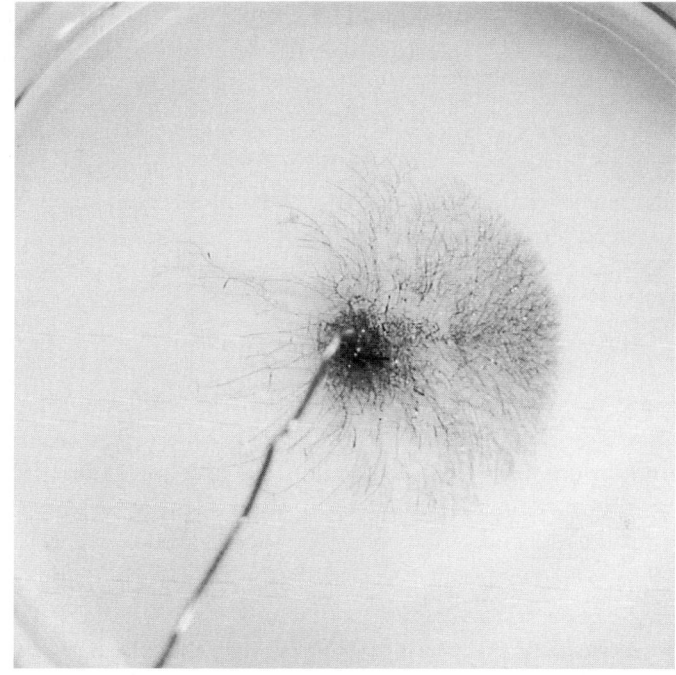

A Lead Tree

Chemicals

Lead acetate, distilled water, acetic acid, zinc rods about 15 cm long and as thick as a pencil (or strips of zinc plate).

Experimental Procedure

20 g of lead acetate are dissolved in about 150 mL of distilled water and made slightly acidic by adding acetic acid. The Erlenmeyer flask containing the clear solution is placed on the base of the stand. A zinc rod (or a strip of zinc plate) is fastened to the stand by means of a double boss; it must dip deeply into the solution. The experiment must be carried out in a pace which is free from vibrations!

The zinc is rapidly covered with a velvety grey coating, from which after a while shiny silver needles grow in a more or less horizontal direction. These build up into skeletal crystals, the lead tree. The spectacle of crystal growth is particularly attractive to the observer when the room is partially darkened and the flask illuminated from the side by a strong light. The experiment takes about half an hour.[2]

A beautiful lead tree is formed much more quickly if ca. 100 mL of a 1.0 mol/L aqueous lead nitrate solution are electrolysed in a crystallising dish using platinum electrodes. With the help of an overhead projector it is easily possible to see how the lead forms a bizarre structure at the cathode in only 2 minutes.

Explanation

The nobler element is deposited onto the zinc. The reduction to metallic lead can also occur electrolytically.

$$Pb^{2+} + 2e^- \rightarrow Pb$$

Waste Disposal

The metallic lead is dissolved in concentrated nitric acid (well-ventilated hood) and the excess acid boiled off. The Pb^{2+} ions are then precipitated as lead sulfate by adding ammonium or sodium sulfate. The precipitate is filtered off and transferred to the breakproof vessel used for collecting less toxic inorganic waste. The filtrate is diluted with tap water and flushed down the drain.

References

1 O. Krätz, *Historisch-chemische Versuche*, Aulis-Verlag, Köln, **1987**, 11.
2 F. Bukatsch, O. P. Krätz, G. Probeck, R. J. Schwankner, *So interessant ist Chemie*, Aulis-Verlag, Köln, **1987**, 189.

5

A Silver Mirror

In 1856 *J. Liebig* published the following procedure for coating glass with silver or gold[1] in his *Annalen:*

One dissolves 100 g of molten nitrated silver oxide in 200 ccm of water and introduces as much caustic ammonia as is required to achieve a clear solution. This liquid is then diluted little by little with 45 ccm of a caustic potash solution having a specific gravity of 1.05 or with the same volume of a 1.035 caustic soda solution .... Once all the caustic potash or soda has been introduced, one dilutes the mixture with sufficient water to produce a volume of 1450 ccm. The mixture is then mixed dropwise with a dilute solution of nitrated silver oxide until a permanent dark grey precipitate is formed (not a cloudiness), and finally enough water is added so that a total volume of 1500 ccm of liquid is achieved. Every cubic centimeter therefore contains slightly more than 6.66 mg of nitrated silver oxide, or 4.18 mg of silver. If the silvering solution is to produce a flawless mirror it must contain no free ammonia whatsoever, but must instead be completely saturated with silver oxide ... Immediately prior to the application of this liquid for silvering purposes one mixes it with 1/10 to 1/8 of its volume of a milk sugar [lactose] solution containing one part by weight of milk sugar in 10 parts of water.

The following observations made by *Liebig* must be taken into account when silver is to be used:

1. The surface to be coated must be scrupulously cleaned; otherwise it will acquire spots;
2. the glass surface must be equidistant from the base of the vessel so that the level of the liquid is everywhere the same and the silver deposition is uniform;
3. the glass surface must be completely wetted by the silvering liquid, and should be washed with alcohol in advance to assure that this will occur more satisfactorily; ...

Apparatus

100-mL wide-necked round-bottomed flask, gas burner, safety glasses, protective gloves.

Chemicals

Silver nitrate $AgNO_3$, hydrazinium sulfate $N_2H_6SO_4$, aqueous ammonia.

Experimental Procedure

A few mL of 0.1 mol/L $AgNO_3$ solution are placed in the flask, which must be completely free of grease. Aqueous ammonia is added until the white precipitate of Ag_2O formed initially just redissolves. A large dash of saturated hydrazinium sulfate solution is added; the flask is swirled rapidly and at the same time heated a little in the gas flame. The silver mirror is deposited on the inner wall of the flask (colored figure 4).

Explanation

The Ag^+ ions are precipitated from the alkaline solution as Ag_2O and redissolve as the diammine silver complex. The hydrazine reduces the silver ions to metallic silver.

$$2\,Ag^+ + 2\,OH^- \rightarrow Ag_2O + H_2O$$
$$Ag_2O + 4\,NH_3 + H_2O \rightarrow 2\,[Ag(NH_3)_2]^+ + 2\,OH^-$$
$$N_2H_6^{2+} + 2\,OH^- \rightarrow H_2O + N_2H_4 \cdot H_2O$$
$$N_2H_4 \cdot H_2O + 4\,Ag^+ + 4\,OH^- \rightarrow N_2 + 5\,H_2O + 4\,Ag$$

Waste Disposal (well-ventilated hood)

Do not pour silver residues down the drain! Silver and silver compounds kill microorganisms in sewage treatment plants. The silver and silver residues are collected, dissolved in concentrated nitric acid and treated with hydrochloric acid (1:1). The AgCl precipitate is allowed to settle, the mother liquor decanted off and filtered, the precipitate again treated with hydrochloric acid (1:1) in a beaker, and the Ag^+ ions reduced by means of zinc rods. The silver precipitate formed is filtered, washed free of chloride ions and zinc and either fused with borax or converted to $AgNO_3$.

Reference

1 J. Liebig, *Liebigs Ann. Chem.*, **1856**, *90*, 132.

6

A Copper Mirror

In some sciences the attempt to find a general principle, an ordo, is often just as fruitless as it would be in biology to seek a general principle or a primordial particle that could have given rise to all living things. Mother Nature does not create genera and species. She creates individuals. Our nearsightedness forces us to look for similarities in order to keep everything in focus ...

Georg Christoph Lichtenberg (1742–1799)[1]

In ancient times copper was known even before iron. In the book „Die Schule der Chemie" (The School of Chemistry) published in 1881 *Julius Adolph Stöckhardt* writes on copper as follows:

Copper was derived in antiquity primarily from the Island of Cyprus, where copper ore was to be found in abundance; this accounts for the name cuprum or Cyprian metal. When it was later regarded a good idea to confer mythological names on the metals, copper was identified with the patron goddess of Cyprus, Venus, as well as with her symbol. Copper possesses numerous outstanding characteristics, which have transformed it into an exceptionally useful metal; namely:
a) It is malleable and durable and ductile; one can therefore hammer it to a foil that, even when very thin, still holds firmly together.
b) It melts only with difficulty (starting at a heat above 1000 °C); therefore it is exceptionally well suited to equipment that will be exposed to fire, such as kettles, pans, alembics, molding forms, etc.
c) It suffers less from rusting in the air than iron; copper utensils thus have a much longer life than those made of iron. The wood of seagoing vessels is clad with copper, as are the roofs of towers and other buildings.
d) It is relatively hard, and therefore wears out less rapidly in its application for copper engraving plates and pressure rollers.
e) It produces very useful alloys with zinc, tin, and nickel; e.g., brass, pinchbeck, bronze, bell metal, gun metal, German silver, etc.
f) It precipitates from solution under the influence of a galvanic current as a firm, continuous mass; this is the way in which one prepares the galvanoplastic impressions of other objects that have recently become so familiar.

g) With oxygen and various acids it produces insoluble compounds of beautiful green and blue colors, which are often used for artistic purposes.

Although copper possesses no odor, it does confer upon sweaty hands and water in which it has stood for a long time (as in copper alembics or kettles) a peculiar, unpleasant odor.

Safety Precautions

Hydrazine hydrate solutions are extremely toxic. The experiment must be carried out in the well-ventilated hood; safety glasses and protective gloves must be worn.

Apparatus

500-mL round-bottomed flask (without a ground glass joint!) or a large test tube, cork ring, gas burner, two measuring cylinders (5-mL and 50-mL), safety glasses, protective gloves.

Chemicals

1 mol/L copper(II) acetate solution, ca. 80 % hydrazine hydrate solution.

Experimental Procedure

30 mL of the 1 mol/L copper(II) acetate solution are introduced into the well-cleaned round-bottomed flask, followed by 5 mL of the hydrazine hydrate solution. The blue solution immediately turns to a brownish green. The copper mirror is formed when the flask is rotated slowly while being heated gently in the gas flame (colored figure 4).

Only 20 mL of the copper(II) acetate solution are required when a large test tube is used.

Explanation

Aqueous hydrazine solutions are strong reducing agents. In the presence of oxidising agents they are thus converted to nitrogen, ammonia or azide (N_3^-) according to equations (1) to (3) below.

$$(1) \quad N_2H_4 \quad \rightarrow \quad N_2 + 4\,H^+ + 4\,e^-$$
$$(2)\; 2\,N_2H_4 \quad \rightarrow \quad NH_4N_3 + 4\,H^+ + 4\,e^-$$
$$(3)\; 2\,N_2H_4 \quad \rightarrow \quad N_2 + 2\,NH_3 + 2\,H^+ + 2\,e^-$$

The relative importance of these three reactions depends highly on the pH value of the solution, the temperature, the oxidising agent used and the catalytic impurities present.

The process described by equation (3) is probably the one mainly involved in the reduction of the copper ions (Cu^{2+}).

$$Cu^{2+} + 2\ e^- \rightarrow Cu$$

Waste Disposal

The reaction solution is collected in the container used for heavy metal solutions.

Reference

1 Cited by F. Cramer, *Chaos and Order*, VCH Weinheim, **1993**, 2.

7

A Mercury Mirror

... they enjoyed employing figures of speech and communicating in symbols and analogies in order that they might be understood only by the prudent, the pious, and the enlightened.

Synesius

The Arab *Abu Mussah Dschafar al Sofi*, also known as *Geber*, taught in the 8th Century that all metals contained mercury and sulfur. It was thus assumed that gold consisted of a large amount of mercury and a small amount of sulfur, whereas copper was thought to contain the two elements in equal measure. In the 12th century the alchemists taught that a prerequisite for the preparation of metals was the philosophers stone, described as follows by the Arab *Khalid*:

This stone combines within itself all the colors. It is white, red yellow, sky-blue, and green.

The Chinese believed that artificial gold was endowed with great magical powers. The masters of the Orient attempted to prepare alloys resembling gold from cinnabar (the naturally occurring form of mercury(II) sulfide) and other metals. They assumed one should eat regularly only from vessels made of such alloys in order to achieve immortality.

Safety Precautions

Mercury and its compounds are highly toxic. Air saturated with mercury vapor contains about 15 mg Hg per m^3! In badly ventilated laboratories the amounts of Hg vapor derived from spilt mercury in the air may suffice to induce chronic mercury poisoning. The inhaled mercury is only very slowly excreted via the urine. Dimethylmercury $(CH_3)_2Hg$ is extremely toxic and can cause irreversible damage to the central nervous system. Ozone is a strong oxidising agent and very toxic.

Apparatus

500-mL round-bottomed flask with a 29/42 ground glass stopper, cork ring, ozone generator, safety glasses, protective gloves.

Chemicals

Mercury, ozone.

Experimental Procedure

Approximately 5 mL of mercury are placed in the round-bottomed flask. The flask is rotated slowly to show that the mercury does not stick to the walls. Ozone is now allowed to flow into the flask for about 1 minute and the flask is stoppered. If the flask is now again rotated slowly the mercury adheres to the glass and forms a mirror.

If the flask is left open for a while on completion of the experiment, the mirror disappears and the mercury runs together to form a globule.

Explanation

The ozone causes a thin oxide layer to form at the surface of the mercury; this lowers the surface tension of the mercury so that it can wet the wall of the flask.

Waste Disposal

Mercury can be resorbed using products such as Chemizorb® (Merck) and placed in a plastic container, the latter should be clearly labelled "Poison".

8

Pharaoh's Snake

We describe one class of existing things as substance, and this we subdivide into three: (1) matter, which in itself is not an individual thing; (2) shape or form, in virtue of which individuality is directly attributed, and (3) the compound of the two. Matter is potentiality, while form is realization or actuality.

Aristotle, "On the Soul"

Friedrich Wöhler (1800–1882) invented the trick known as Pharaoh's Snake, which became very popular at fairs:

It consists of a cone of pressed gray-white powder, which is ignited on a fire-resistant surface. Bluish flames, out of which arose a fantastically shaped monster of a brownish-yellow color that was well suited to test one's memories of the Biblical story according to which Moses and his companions cast rods in front of the Pharaoh and transformed them into snakes ...

This description of the experiment was provided by the dyestuffs chemist O. N. Witt.[1]

The light grey powder was $Hg(SCN)_2$, which is highly toxic. We have thus devised a less dangerous form of the experiment without loosing the beauty of the effect.

Apparatus

Fireproof support or porcelain dish, safety glasses, protective gloves.

Chemicals

Sand, ethanol, pellets of $NaHCO_3$ and icing sugar.

Experimental Procedure

A cone of sand is formed on the fireproof support or in a porcelain dish. Three pellets are placed on top of the cone, at least 5 mL of ethanol are added and light is put to it. After the alcohol has burnt off a black porous mass emerges

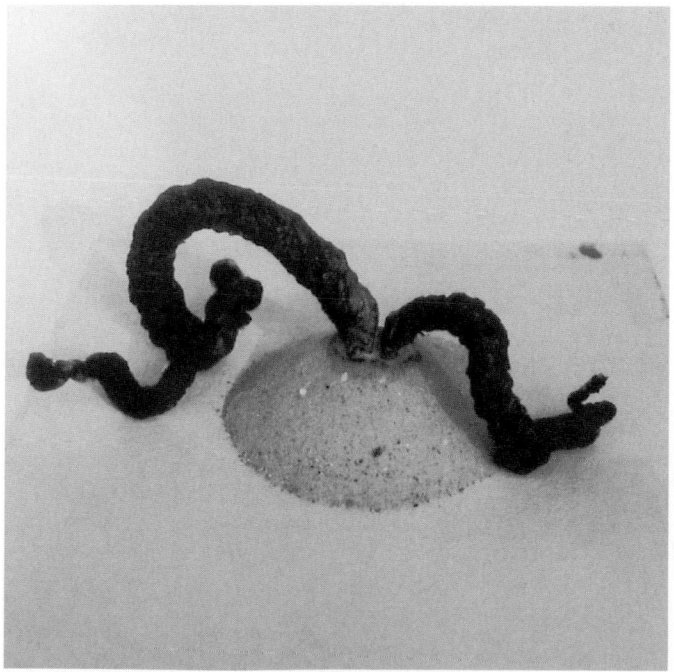

Pharaoh's Snake; at the top: before lighting, below: after lighting.

from the "volcano cone"; it resembles a snake and grows ever larger. It is about as thick as a finger and can be up to 1 m long.

Explanation

The melted sugar and the gas formed when the sodium bicarbonate is heated produce an extremely voluminous foamy mass. A part of the sugar ultimately burns, undergoing partial carbonization. The mixture of the decomposed salt and carbon forms the snakelike mass.

Waste Disposal

Residues can be disposed of with the household waste.

Reference

1 Cited by O. Krätz, *Historisch-chemische Versuche,* Aulis-Verlag, Köln, **1987,** 63.

9

A Chemical Garden

Thomas Mann beautifully combines sensory observation and fantasy in his description of a chemical garden in the book *"Doctor Faustus"* which he wrote in his old age:

> *I shall never forget the sight. The vessel of crystallization was three-quarters full of slightly muddy water – that is, dilute waterglass – and from the sandy bottom there strove upwards a grotesque little landscape of variously colored growths: a confused vegetation of blue, green, and brown shoots which reminded one of algae, mushrooms, attached polyps, also moss, then mussels, fruit pods, little trees or twigs from trees, here, and there of limbs. It was the most remarkable sight I ever saw, and remarkable not so much for its appearance, strange and amazing though that was, as on account of its profoundly melancholy nature. For when Father Leverkühn asked us what we thought of it and we timidly answered him that they might be plants: 'no,' he replied, 'they are not, they only act that way. But do not think the less of them. Precisely because they do, because they try as hard as they can, they are worthy of all respect'* ...

Apparatus

1500-mL or 2000-mL beaker, watchglass to cover the beaker, safety glasses, protective gloves.

Chemicals

500 mL commercial waterglass (an aqueous solution of sodium silicate), distilled water, crystals (as large as possible) of the following hydrated salts, the colors which these form at the membrane being given in brackets: $AlCl_3$ (white), $CoCl_2$ (dark blue), $CrCl_3$ (dark green), $CuCl_2$ (light blue-green), $FeCl_3$ (yellow and brown), $MnCl_2$ (white and pale pink), $CaCl_2$ (white), $Ni(NO_3)_2$ (green). $KMnO_4$ (violet) is also suitable.

Experimental Procedure

The waterglass solution is diluted with an equal amount of distilled water and about a liter of the resulting solution is poured into the beaker. The crystals of the various salt hydrates are now distributed in the solution in such a manner that the bottom of the beaker is evenly covered. The watchglass is now placed over the beaker. After a short while a colorful garden with lush vegetation is formed (colored figure 5).

Explanation

The waterglass solution reacts with the metal ions to form a semipermeable membrane consisting of an almost insoluble precipitate of metal salts. Since the concentration of the dissolved metal salts is greater in the space between crystal and membrane than in the surrounding solution water diffuses into this space. The osmotic pressure thus increases and the membrane expands or bursts. The hole thus formed is immediately filled by the metal salt. The salt concentration is lowest at the highest point of the membrane, so that the latter normally bursts here and the "plants" grow upwards as in nature.

Waste Disposal

On completion of the experiment the complete chemical garden is treated with dilute milk of lime, the pH being kept below 8. The mixture is then centrifuged (or simply decanted), separated from the precipitated mixture and the filtrate diluted and poured down the drain. The residue is placed in the plastic container used for collecting less toxic inorganic waste.

10

Self-Igniting Iron

I happened to be in a forge towards evening at the moment when a glowing mass of iron was placed on the anvil; I had fixed my eyes stead-fastly on it, and, turning round, I looked accidentally into an open coal-shed: a large red image now floated before my eyes, and, as I turned them from the dark opening to the light boards of which the shed was constructed, the image appeared half green, half red, according as it had a lighter or darker ground behind it. I did not at that time take notice of the subsequent changes of this appearance.

Johann Wolfgang von Goethe, "Theory of Colors" (1810)

The best-known corrosion process is the rusting of iron. Luckily this process, which economically speaking is a disaster, occurs only slowly. As is well known, it is due to the conversion of iron atoms fixed in the crystal lattice into hydrated iron(III) oxide, the "rust", in the presence of air and water. If the stable lattice is destroyed, the oxygen molecules of the air can react freely with the finely-divided iron particles, so that the corrosion process appears to occur without any energy being supplied.

Apparatus

4 wide-necked test tubes with rubber bungs, fire-resistant support, gas burner, safety glasses, protective gloves.

Chemicals

Iron oxalate, steel wool or filings.

Experimental Procedure

The four open test tubes are filled to about 1/5 of their height with yellow iron oxalate, $FeC_2O_4 \cdot 1.5 H_2O$ and each tube is heated until the latter has been converted to a black residue. Any water vapor which condenses at the tops of the tubes is carefully removed with a soft filter paper. The test tubes are then tightly closed with the rubber bungs and put aside for the demonstration. After showing the auditorium how quickly steel wool burns with a bright glow in the flame of the gas burner, the flame is extinguished and the room darkened.

The lecturer and his assistant (or two spectators) now stand on the demonstrating table, hold the test tubes upside down and remove the bungs simultaneously. A brilliant shower of red sparks demonstrates that the iron oxidises spontaneously.

Explanation

Iron oxalate decomposes on heating to give a pyrophoric mixture of iron(II) oxide and metallic iron, which reacts with oxygen according to the highly exothermic reaction described by the following equation:

$$Fe + FeO + O_2 \rightarrow Fe_2O_3$$

Waste Disposal

The solid residue is transferred to the container used for collecting less toxic inorganic substances.

11

The Passivation of Iron

I have made you an assayer of my people
– A refiner –
You are to note and assay their ways.
They are bronze and iron
They are all stubbornly defiant;
They deal basely
All of them act corruptly.
The bellows puff;
The lead is consumed by fire.
Yet the smelter smelts to no purpose –
The dross is not separated out.
They are called "rejected silver",
For the Lord has rejected them.

The Book of Jeremiah 6, 27–30

In *Hammurabi's* Babylon iron was the next most expensive element after silver: two shekels of silver cost eight of iron and 120–140 shekels of copper. The iron column near Delhi is more than 1500 years old, is 7.66 m in height and weighs 6 t. It consists of 99.72 % pure iron (as well as traces of C, Mn, S and P) and has retained its purity throughout the centuries. And it is symbolic that the Atomium built in Brussels in 1958 consists of nine iron spheres which represent the cubic body-centered lattice structure of the stable modification α-iron.

Apparatus

Two glass vessels (10 × 10 × 5 cm), spatula, piece of sheet iron fitted with a handle, safety glasses, protective gloves.

Chemicals

Nitric acid, 5 % copper nitrate solution.

Experimental Procedure

The sheet iron is immersed in the concentrated nitric acid contained in the first glass vessel, and after about 10 seconds transferred to the second glass vessel which contains the copper nitrate solution; there is no visible change at the surface. If however the iron surface is damaged, for example by scratching it with the spatula, a layer of copper rapidly covers the surface. The same process also occurs when the iron is treated first with dilute nitric acid and then with the copper nitrate solution. The passivation process can be repeated several times.

Explanation

The treatment of the iron with concentrated nitric acid HNO_3 (or chromic acid, H_2CrO_4) leads to the formation of an oxide layer which covers the whole surface of the metal; this layer prevents the diffusion of oxygen and moisture from the air into the iron lattice. However, if the surface is untreated, corrosion sets in. This is greatly favored by the high cell voltage which is built up between the base metal iron and the hydrated copper ions Cu^{2+} or when the iron is treated with dilute acids (H_3O^+ ions); as a result Fe^{2+} ions go into solution.

(1) $Fe + Cu^{2+}_{aq} \rightarrow Cu + Fe^{2+}_{aq}$ ($\Delta E° = 0.78$ V)
(2) $Fe + 2\ H_3O^+ \rightarrow H_2 + Fe^{2+}_{aq}$ ($\Delta E° = 0.44$ V)

Waste Disposal

The solutions should be transferred to the container used for the collection of less toxic inorganic waste.

12

Fireworks Ignited by Ice

How strangely, thro' each dark ravine,
A dim light blushes like the morn!
Pervading every rift 'tis seen,
That in the mountains's side is torn!
Black fogs and gauze-like vapours mount,
Thro' which the streaming glow is spread,
Now creeping like a slender thread,
Now gushing up, a fiery fount!
Its downward course it now maintains,
Meandering thro' a hundred veins,
Till, in too close a channel pent,
Exploding, it in foam is sprent;
And fly the globules far and wide,
Like golden sand on every side!
And look! the mountains's steepy height,
From base to summit, burns outright!

Johann Wolfang von Goethe, "Faust, The First Part",
May-Day Night, Faust speaking to Mephistopheles

Safety Precautions

Wear safety glasses!

Apparatus

Ceramic wire gauze, crucible tongs, glass rod, spatula, two small glass vessels (5-mL and 25-mL), filter paper, safety glasses, protective gloves.

Chemicals

Zinc dust, ammonium nitrate, ammonium chloride, ice cubes, barium nitrate.

Experimental Procedure

4 g of zinc dust are placed in the 5 mL vessel, while a mixture of 4 g of NH_4NO_3, 1 g of NH_4Cl and 0.5 g of $Ba(NO_3)_2$ is prepared in the larger one.

The contents of the two vessels are carefully mixed on a piece of filter paper (avoid friction!) and formed into a cone. The reaction is "ignited" by one or two small ice cubes, which are carefully placed on the cone using the crucible tongs. After a few seconds a pale green jet of spark-emitting flame (barium!) shoots up from the cone (colored figure 6).

Explanation

The water film on the surface of the ice cubes initiates the strongly exothermic redox reaction between the zinc dust and the nitrate (which supplies oxygen); ammonium chloride accelerates the process and itself evaporates (smell of NH_3!). The zinc is oxidised to ZnO.

Waste Disposal

The residue is transferred to the container used for less toxic inorganic residues.

13

The Aluminum-Iodine Reaction

It is easy to work, wide-spread, for almost every stone contains it, it is three times lighter than iron and it seems that it has been specially created for making projectiles.

Jules Verne, *"Traveling to the Moon" (1865)*

Verne is referring to aluminum, that "silver from clay" which was first obtained in a pure state 1827 by *Wöhler* from the reduction of aluminum trichloride with potassium. The name of the newly discovered element comes from the Latin word "alumen", which refers to the alums, minerals containing aluminum which were used for medicinal purposes even in ancient times. In 1855 this metal, whose properties are praised by *Jules Verne*, was shown next to the Crown Jewels at the Paris World Exhibition, while *Napoleon III* used aluminum cutlery at state dinners. Today materials science would be inconceivable without the use of aluminum, and since the turn of the century its price in the world market has remained below a dollar per kg. The extreme reactivity of this light metal, in particular towards electronegative partners, can only be observed when the stable oxide layer at the metal surface is destroyed or the metal is in a finely divided form. Iodine is a particularly interesting reaction partner, which can easily be manipulated because it is a solid.

Safety Precautions

The reaction should be carried out in a well-ventilated hood. Wear safety glasses and protective gloves!

Apparatus

Fire-resistant support, 100-mL beaker, glass rod, dropping pipette, safety glasses, protective gloves.

Chemicals

Pulverised iodine, aluminum powder, distilled water.

Experimental Procedure

12 g of finely powdered iodine are mixed well in a beaker with 2 g of aluminum powder and the mixture formed into a cone on the fire-resistant support. On the addition of 2–3 drops of water a cloud of smoke is immediately formed, and a few seconds later the mixture burns with a violet flame which is darkened by the iodine vapor. The bright violet color is soon transformed to a fine dark brown powder which covers everything.

In the same way it is possible using the same relative amounts to bring the element magnesium, which was obtained for the first time in a pure state by *Humphry Davy* in 1808, to reaction with iodine.

Explanation

The reaction between solid iodine and solid aluminum is induced by water, the enthalpy which is set free at the beginning of the reaction sufficing to convert the whole mixture to dialuminum hexaiodide and to sublime the excess iodine. Pure iodine exhibits its distinctive violet color, which changes to a dark brown on contact with oxygen-bearing substances such as water, paper etc. The relatively low dissociation enthalpy of the iodine molecule ($\Delta H^0 = +151$ kJ/mol) favors its cleavage to give iodine atoms, which in turn react with aluminum to give Al_2I_6:

$$2\ Al + 3\ I_2 \rightarrow Al_2I_6$$

The reaction of magnesium with iodine to give MgI_2 occurs in a similar manner.

Waste Disposal

The solid residue is washed with a dilute soda solution. The liquid is decanted from the precipitate; the latter is transferred to the container used for collecting less toxic inorganic waste, while the solution can be flushed down the drain.

14

Lightning under Water

"Close to the psalter there lay an obviously just finished exquisite golden booklet which was so incredibly small, that one could have taken it in the flat of the hand. At first sight the miniatures next to the tiny characters were scarcely to be seen and required to be looked at very closely so that all their beauty could appear (and you would ask yourself what kind of superhuman instrument the artist might have used to obtain such an vivacious expression in such a reduced space.) The entire margins of the book were covered with tiny little figures which were generated as by natural expansion out of the final arches of the marvelously painted characters: marine sirens, fleeing deer, chimera, human torsi without arms coming out of the end of the verses like worms. At one place, kind of continuation of the three "Sanctus, Sanctus, Sanctus" which were repeated in three different lines, I saw three animal figures with human heads, of which two bowed, one to the bottom one to the top in order to unite in a kiss which you would not hesitate to call shameless if you would not be convinced that unquestionably this representation at that place was certainly justified by some profound but not easily recognizable spiritual signification."

Umberto Eco "The name of the rose"

Thus the Italian poet and philosopher *Umberto Eco* describes the *"World of the smallest"* in his novel *"The name of the rose"*.

Safety Precautions

Dimanganese heptoxide oxidises organic substances such as alcohols explosively at temperatures of about 100 °C. Great care must therefore be taken that the system does not heat up prior to the introduction of the potassium permanganate.

Apparatus

Two test tubes (16 × 160 mm), two stands with fixing devices, 200-mL beaker, pneumatic through, pipette, safety glass, protective gloves.

Chemicals

Potassium permanganate $KMnO_4$, concentrated sulfuric acid, ethanol or *n*-propanol.

Experimental Procedure

The two test tubes are fixed vertically to the stand in such a way that about half of their volume is immersed in the trough, which is filled with tap water. They are now filled to a height of about 2 cm with concentrated sulfuric acid, making sure that the acid does not come in contact with the top edge of the tubes. A layer of alcohol 4 cm in height is now allowed to flow carefully on to the surface of the sulfuric acid using a pipette. Great care must be taken to make sure that the liquids do not mix, since this will cause a strongly exothermic reaction to occur! A $KMnO_4$ crystal about 3 mm long is now dropped into each test tube, and shortly afterwards yellow, lightning-like sparks can be seen at the interface between the liquids; these are accompanied by crackling noises. The sparks appear to be formed "under water". The reaction mixture slowly turns green and in some places brown as a result of the redox processes taking place. The experiment can take up to 15 min, the appearance of the "lightning" being completely unpredictable (colored figure 7).

Explanation

$KMnO_4$ reacts primarily with the sulfuric acid to give dimanganese heptoxide;[1] the latter oxidises the alcohol in a highly exothermic reaction and itself is converted to manganese dioxide MnO_2.

$$2 \ KMnO_4 + H_2SO_4 \rightarrow K_2SO_4 + Mn_2O_7 + H_2O$$
$$Mn_2O_7 \rightarrow 2 \ MnO_2 + 1.5 \ O_2$$

Waste Disposal

After cooling, the contents of the test tubes should be poured into a beaker of cold water (safety glasses!). The mixture is neutralised with milk of lime and the precipitate transferred to the container used for storing less toxic inorganic salts. The solution is poured down the drain.

Reference

1 M. Trömel, M. Russ, *Angew. Chem.*, **1987**, *99*, 1037; *Angew. Chem. Int. Ed. Engl.* **1987**, *26*, 1007.

15

Growling Gummy Bear

He who ponders long over four things were better never to have been born: that which is above, that which is below, that which came before, and that which comes hereafter.

Talmud

Berchtold, Duke of Zähringen, is said to have proclaimed as he founded the city of Bern in Switzerland:

"... just as the bear is the largest and mightiest animal in the land, so will I make powerful the city named in his honor."

The bear appears as a symbol of valor on the arms of the chivalrous Order of the Bear, founded by *Emperor Friedrich II* in the year 1213 and endowed by him out of gratitude to the loyal subjects who had stood by his side in driving *Otto IV* out of the empire.

Safety Precautions

The reaction must be carried out behind a safety screen or in a well-ventilated hood. No flammable substances should be located near the experiment, and the support must be fire-resistant.

Potassium chlorate is a very potent oxidising agent which reacts with most organic substances so violently that the reaction mixture can catch fire or even explode.

Apparatus

Large test tube, gas burner, stand, clamp and boss, safety glasses, protective gloves.

Chemicals

Potassium chlorate, gummy bears.

Experimental Procedure

10 g of potassium chlorate are placed in the test tube and fused using the gas flame. A gummy bear is then added. The gummy bear burns brightly, "dances"

on the molten salt and emits a curious "growling" noise (colored figure 8). **Take care!** The reaction is often so violent that potassium chlorate is shot out of the test tube by the carbon dioxide and water evolved. The test tube should thus be fastened at a slight angle and **under no circumstances should it point at the audience!**[1]

Waste Disposal

The excess potassium chlorate is boiled with aqueous hydrochloric acid and then neutralised with sodium hydroxide. The salt solution can then be poured down the drain.

Reference

1 D. M. Sullivan, *J. Chem. Educ.*, **1992**, *69*, 326.

16

Self-Igniting Wood Wool

Yea, for the king it is made ready; he hath made it deep and large: the pile thereof is fire and much wood; the breath of the Lord, *like a stream of brimstone, doth kindle it.*

The Book of Isaiah 30, 33

From the earliest times fire has had a great symbolic power for Man, and just like air, earth and water, which can extinguish even the greatest conflagrations, it has remained one of the elements which has determined the actions of Man and Nature. Here we shall show (without resorting to magic!) how water itself can start a fire.

Apparatus

One-liter beaker, fire-resistant support, safety glasses, protective gloves.

Chemicals

Wood wool or wood shavings, powdered sodium peroxide, mineral or tap water.

Experimental Procedure

Prior to the demonstration a quantity of wood wool or wood shavings is placed in the beaker and covered with a layer of powdered sodium peroxide. The experimentalist takes a drink of the water he will use to "ignite" the wood to show the public that it is normal water and then pours a little of it over the contents of the beaker. The wood shavings very quickly start to burn and in a few seconds they are converted to ash.

Explanation

The thermally relatively stable sodium peroxide Na_2O_2 reacts with water in a strongly exothermic reaction to give sodium hydroxide and hydrogen peroxide; the heat evolved causes the wooden shavings to ignite.

(1) $Na_2O_2 + 2 H_2O \rightarrow 2 NaOH + H_2O_2$
(2) $H_2O_2 \rightarrow H_2O + 0.5 O_2$
(3) $Na_2O_2 + CO_2 \rightarrow Na_2CO_3 + 0.5 O_2$

The hydrogen peroxide formed disproportionates quickly under the influence of heat into water and oxygen, which sustains and encourages the combustion process.

As wood shavings have a high surface area, the reaction is extremely fast.

It should be mentioned that the French chemist *L. J. Thénard* was the first to obtain hydrogen peroxide (in 1818) by reacting BaO_2 with dilute mineral acid. Today Na_2O_2 is used in breathing apparatus (for example by fire brigades) and in submarines for CO_2/O_2 exchange according to equation (3).

17

Fire without Matches

On the evening of the 13th of June 1847 heavy smoke was seen pouring out of the windows of a house (number 17) in the Neckarstraße in Darmstadt. The doors were broken open and the body of the Countess of *Görlitz* was found lying half-burned in one of the rooms.

Justus Liebig and *Heinrich Emanuel Merck* were called as expert witnesses in the trial that followed. The main question was as to whether it is possible for a human body to catch fire on its own. *Liebig* considered this to be impossible. It was finally established that her servant had murdered the lady, robbed her and burnt the body. The main witness was *Friedrich August Kekulé*, who was one of the first to enter the smoke-filled room. After the trial he studied chemistry in Gießen and took his doctorate there in 1852. In Heidelberg he recognised that the four bonds formed by carbon are equivalent (as the Scottish chemist *A. S. Couper* discovered independently at about the same time) and that the element can form chains containing C-C bonds. In Ghent, where he was a professor from 1858 to 1867, *Kekulé* postulated the ring structure of benzene in 1865, one of the great theoretical achievements of the time.

But back to the criminal case! We shall show that it is indeed possible for substances to undergo self-ignition.

Apparatus

Mortar and pestle, spatula, two dropping pipettes, two evaporating dishes, safety glasses, protective gloves.

Chemicals

10 g Potassium permanganate, 5 ml glycerine.

Experimental Procedure

5 g of $KMnO_4$ crystals and 5 g of finely powdered potassium permanganate are formed into little cones in the two evaporating dishes; a small hollow is formed in each cone using the spatula. Two mL of glycerine are now dropped carefully and simultaneously into each hollow using the dropping pipettes. After only a few seconds white smoke is formed, followed by the formation of flying sparks accompanied by decrepitation and a purple flame; the reaction is very violent

and clearly occurs more intensely and rapidly in the dish containing the powdered salt.

Explanation

The oxidation of the glycerine by potassium permanganate is initially slow, but accelerates as the system becomes hotter and eventually catches fire. The small amount of green residue probably consists of K_2MnO_4, the dark brown or black residue is a mixture of MnO_2 and Mn_2O_3, while the white product is K_2CO_3 which is formed from CO_2, the oxidation product formed from the glycerine. The higher rate at which the powdered $KMnO_4$ reacts is due to its greater surface area.

Waste Disposal

The residue is stirred with water and left to stand for a while. The solution is then decanted off from the solid, which is transferred to the container used for storing less toxic inorganic waste. The solution is poured down the drain.

18

White Phosphorus, a Very Dangerous Chemical

L. *Hofmann* sent from Jamaica the following letter to the editors of the journal *Annalen der Chemie und Pharmazie* regarding a new reaction for detecting phosphorus in cases of poisoning:[1]

In November 1861 I was presented for chemical investigation the entrails of four children siblings one of whom, an infant, died suddenly in minutes, the others equally suddenly within 24 hours, all with symptoms suggestive of poisoning. Fourteen days after the deaths I received the entrails in clay pots sealed with cork stoppers wrapped in cotton towels. As these were being transported by sea to Kingston they were exposed for many days to the tropical sun, a circumstance which I could hardly overlook. Upon opening the vessels I was struck by microscopic crystals that had collected on the fibers of the cloth, and which proved upon closer investigation to be phosphate of ammonia.

In order to discover the phosphorus I subjected a portion of the solid entrails, together with liquid from the vessel and after the addition of some oxalic acid, to a distillation in the dark; but I was unable to detect the least bit of luminescence in the air in the retort and the receiver.

A portion of the distillate was treated with a little sulphurous ammonia [ammonium hydrogensulfide] and evaporated as a test for prussic acid; a trace of residue remained, which I attempted to convert to sulphurous prussic acid by treatment with iron chloride; to my great astonishment I observed a violet reaction, which rapidly disappeared, however, without a trace of blood-red.

Meanwhile, I had received from another part of the island the entrails of another person who had died under suspicious circumstances. The liquid isolated from these entrails also provided me with no trace of phosphorescence, but it again displayed quite strikingly the blue reaction in the course of our procedure for the detection of prussic acid.

I thought at first of a laudanum poisoning, and an overdose of morphine or meconic acid, but close investigation showed that the distillate was completely devoid of these substances. The violet reaction must therefore neces-

sarily be derived from a volatile material, which in combination with sulphur or sulphurous ammonia brought about the dark-violet reaction with iron chloride. Evaporation of my sulphurous ammonia left no residue that might have been associated with this reaction. I came then upon the idea that a sulphur phosphorus compound was the cause of this coloration with iron chloride, since I was aware that many circumstances prevent the luminescence of phosphorus; and, indeed, when I treated with a little sulphurous ammonia and then evaporated about two drachmas of water in which phosphorus had been stored, a drop of iron chloride solution added to the dry residue produced a very distinct fleeting coloration, namely violet. A very small piece of phosphorus boiled with sulphurous ammonia and then distilled gave the same reaction. Distillation of a somewhat putrefied brain together with a match, and evaporation of the distillate with sulphurous ammonia, led to a violet coloration with iron chloride which then became dark brown.

So much for this precise description. Shortly after his discovery of phosphorus in 1669, the alchemist *Hennig Brand* himself, the person most intimately acquainted with the material, still handled the new substance quite carelessly, After all, a lethal oral dose of white phosphorus amounts to only 50 mg. At that time apparently neither the substance's toxicity nor its extreme tendency toward self-combustion was recognized. *Brand* himself came to experience both phenomena. In a letter to *Gottfried Wilhelm Leibniz* on April 30, 1679, he wrote:

When in these days I had some of that very fire in my hand and did nothing more than blow on it with my breath, the fire ignited itself as God is my witness; the skin of my hand was burned truly into a hardened stone such that my children cried and declared that it was horrible to witness.[2]

With appropriate precautions it is possible to demonstrate safely several experiments that convey in an impressive way the interesting characteristics peculiar to the white modification (P_4) of phosphorus.

Safety Precautions

White phosphorus is highly toxic and is pyrophoric even at room temperature. Safety glasses and protective gloves must be worn!

Apparatus

Two 600-mL beakers, two large test tubes, inlet tube with pointed tip, about 25 cm in length, crucible tongs, glass rod, tripod, wire gauze, copper sheet

(3 × 20 cm, 1–2 mm thick), gas burner, stand with fixing device, pincette, safety glasses, protective gloves.

Chemicals

White and red phosphorus, candle with a long wick, carbon disulfide, distilled water, filter paper.

Experimental Procedure

(I) The beaker is placed on the wire gauze and the test tube hung inside the beaker. Both are filled with water to the same level. A piece of white phosphorus the size of a pea is placed in the test tube and the water bath warmed. The phosphorus begins to melt when the temperature of the bath reaches about 50 °C. If a stream of oxygen is now passed into the test tube the phosphorus at once begins to burn.

(II) The sheet of copper is placed on a tripod. A small piece of red phosphorus is placed on one end of it and a piece of white phosphorus (previously dried with a filter paper) on the other. The gas burner is placed under the mid-point of the copper sheet and lit. The white phosphorus starts to burn almost at once, while the red modification ignites only at a much higher temperature; in both cases white smoke is formed.

(III) A candlewick, which should be long, loosely woven and previously been cleaned, is impregnated with a solution of P_4 in CS_2, prepared by dissolving 0.1 g of white phosphorus in 2 mL of carbon disulfide. Care must be taken that no naked flames are in the vicinity! The candle is now placed in a beaker and the room darkened; after a while the candle begins to burn, and the luminous flame typical of phosphorus is observed. The self-ignition of white phosphorus in air can also be demonstrated by dipping a strip of filter paper into the solution of white phosphorus in CS_2. The filter paper is placed on a fire-resistant support using a pair of crucible tongs; as soon as it has dried it begins to burn.

Explanation

The experiments I to III demonstrate the pyrophoric nature of the white modification of phosphorus (melting point 44.1 °C). The reactivity of P_4 is mainly due to the extremely small bond angles in the tetrahedral P_4 molecule, which is thus subject to considerable ring strain. White phosphorus is oxidised to P_4O_{10} via various intermediate phosphorus oxides. During the oxidation a pale light emission is observed. When white phosphorus is dissolved in carbon disulfide the P_4 molecules are dispersed in the solvent, and the rapid evaporation of CS_2 (boiling point 46.2 °C) is also accompanied by the oxidation of the phosphorus. Red phosphorus, in contrast, does not melt until about 600 °C and is unreactive under the conditions used for the above experiments because of its polymeric structure.

Waste Disposal

Phosphorus residues are treated with concentrated copper sulfate solution to give the corresponding phosphide, which is oxidised by a strongly alkaline sodium hypochlorite solution (cooling is required!); milk of lime is then added to give calcium phosphate and copper hydroxide. These solids are both only very slightly soluble in water and are transferred to the container used for collecting less toxic inorganic waste. The aqueous solution is neutralised with dilute sulfuric acid and poured down the drain.

Reference

1 L. Hofmann, *Ann. Chem. Pharm.*, **1863**, *125*, Neue Reihe, *49*, 121.
2 Cited by H. W. Prinzler, *Phosphorus, Sulfur, Silicon*, **1993**, *78*, 8.

19

Fireworks for a Garden Party –
Red and Green Fire

The German poet *E. T. A. Hoffmann* describes a firework display in his "Lebensansichten des Katers Murr" ("The Philosophy of Murr the Tom Cat") (1819):

> *Soon the court and the entire party were in place. After the usual displays with Catharine wheels, rockets, fire-balls, and other ordinary things, there finally appeared the princess's monogram in the form of Chinese diamond fire; but high above that yet there floated in the air and gradually faded away the name Julia in milky-white light. Now it was time. I ignited the girandole, and as the rockets hissed and crackled into the heavens, the weather broke loose with fiery red lightning, with crashing thunder, causing the forest and mountains to resound. And the tempest raged into the park and aroused a howling cry from a thousand voices in the deepest thickets. I tore the instrument from the hand of a fleeing trumpeter and blew a lusty triumphant blast, while the artillery salvoes from the firepots, the maroons, and the cannons gallantly countered the explosions from the rolling thunder.*

It was decided that a great firework display should be held in Green Park in April 1749 to celebrate the signing of the Treaty of Aix-la-Chapelle.

Händel was commissioned to compose a "martial overture" to provide a musical background to the fireworks. This grew into the Music for the Royal Fireworks. The king, who had always considered himself to be a soldier, in particular after the victory at Dettingen, insisted on the use of "warlike" instruments such as brass. However, *Händel* could not bring himself to use only brass but wanted to include string instruments. Even today we still do not know whether violins played on that day or not.

A public dress rehearsal was held in Vauxhall Gardens on April 22nd; 12,000 attended, and the innumerable coaches caused the very first and now famous traffic jam in London!

The actual performance with fireworks took place on April 27th in Green Park. Unfortunately the famous "firework machine" and the pavilion which had been designed by Signor *Servandoni* from the Royal Opera House caught fire. The hot-blooded Italian became so angry that he drew his sword against poor Mr. *Frederick* the ordnance officer. Fortunately the Italian was disarmed before he could cause any damage; he was arrested and forced to apologise the next day.

At the end of the day it began to rain (hardly surprising). However, in spite of the problems with the fireworks, *Händel*'s music was a great success and has remained so until today.

Safety precautions

The reaction should on no account be carried out in a closed room, as a great deal of smoke and sulfur dioxide are formed! If the experiment is carried out in well-ventilated hood only 1/10th of the stated amounts of chemicals should be used. The mixture should never be ground in a mortar using a pestle. **Danger! Explosive!** Organic compounds can react explosively with $KClO_3$. Always wear safety glasses and protective gloves!

Apparatus

Sheets of paper, two fire-resistant supports, measuring beaker, safety glasses, protective gloves.

Chemicals

Strontium nitrate $Sr(NO_3)_2$, powdered charcoal, flowers of sulfur, potassium chlorate $KClO_3$, barium nitrate $Ba(NO_3)_2$.

Red Fire

75 g $Sr(NO_3)_2$, 2.7 g powdered charcoal, 32 g flowers of sulfur, 63 g potassium chlorate.

Green Fire

75 g $Ba(NO_3)_2$, 2.7 g powdered charcoal, 32 g flowers of sulfur, 63 g potassium chlorate.

Experimental Procedure

The chemicals are weighed out, transferred to a large sheet of paper and cautiously mixed by moving the paper to and fro. The mixture is placed on a fire-resistant support and ignited using a gas flame or a few drops of concentrated sulfuric acid.

Waste Disposal

The solid residue is stirred with an aqueous soda solution. The filtrate is poured down the drain and the precipitate transferred to the container used for collecting less toxic heavy metal salts.

Red, Green, and Yellow Fire
without Emission of Sulfur Dioxide

In ancient India, in the province of Bengal, the priests pressed mixtures containing strontium salts, charcoal, sulfur and potassium chlorate into balls or pyramides and ignited them in the gloom of the temples: *Bengal Fire*. A magnificent firework display can be obtained even without sulfur. In 1540 the Venetian *Vannocio Biringuccio* wrote in his book *"De la pirotechnia"* that these fascinating displays unfortunately last "only as long as a lover's kiss or perhaps not even that long".

Safety Precautions

Large amounts of smoke are emitted during the reaction! Never mix the chemicals in a mortar! Potassium chlorate and icing sugar can react explosively. Always wear safety glasses and protective gloves.

Apparatus

Large sheets of paper, two fire-resistant supports, dropping pipettes, safety glasses, protective gloves.

Chemicals

Potassium chlorate $KClO_3$, icing sugar, barium nitrate $Ba(NO_3)_2$, strontium nitrate $Sr(NO_3)_2$, concentrated sulfuric acid, sodium nitrate $NaNO_3$.

Red Fire

10 g $KClO_3$, 10 g icing sugar, 20 g $Sr(NO_3)_2$.

Green Fire

10 g $KClO_3$, 10 g icing sugar, 20 g $Ba(NO_3)_2$.

Yellow Fire

10 g $KClO_3$, 10 g icing sugar, 20 g $NaNO_3$.

Experimental Procedure

The chemicals are weighed out, transferred to a large sheet of paper and mixed cautiously by moving the paper to and fro. They are poured on to the fire-resistant to form an elongated heap. The mixtures are ignited at one edge with 1–2 drops of concentrated sulfuric acid added from a dropping pipette; a sparkler can also be used.

Waste Disposal

This is carried out as in the previous experiment.

21

Artificial Fog

That all our knowledge begins with experience there can be no doubt. For how should the faculty of knowledge be called into activity, if not by objects which affect our senses, and which either produce representations by themselves, or rouse the activity of our understanding to compare, to connect, or to separate them; and thus to convert the raw material of our sensuous impressions into a knowledge of objects, which we call experience? In respect of time, therefore, no knowledge within us is antecedent to experience, but all knowledge begins with it.

Immanuel Kant, "Critique of Pure Reason", Introduction (1781)

Safety Precautions

A large amount of smoke is evolved during the reaction, which must therefore be carried out either in the open air or in a well-ventilated hood. Wear safety glasses! KOH can cause severe damage to the skin or eyes, and in case of skin contact the area affected should be washed for at least 15 minutes under running water. The same applies for eye contact, but in this case a doctor should be consulted in addition. The toxicity of hydroxylammonium chloride is unknown, but there is a possibility that it is cancerogenic.

Apparatus

Two 50-mL beakers, medium-size tin can or 250-mL beaker, glass rod, safety glasses, protective gloves.

Chemicals

KOH in pellet form, hydroxylammonium chloride.

Experimental Procedure

20 g of KOH and 20 g hydroxylammonium chloride are weighed out into the two 50-mL beakers. The two chemicals are then mixed rapidly and thoroughly in the tin can with the help of a glass rod. The reaction starts better when the mixture is formed into a heap in the reaction vessel. The reaction starts after about 30 seconds: white smoke is evolved.

Explanation

The smoke consists mainly of water, ammonia, ammonium chloride and nitrogen. However, its composition has still not been exactly determined, and it is possible that highly toxic compounds may be present in small amounts.

Waste Disposal

The residue in the can should be dissolved in water, neutralised using dilute hydrochloric acid and poured down the drain.

22

Mushroom Clouds

Many thousands of years ago, shepherds in the Near East and Egypt are said to have isolated a white readily sublimable substance from camel dung and salt; later this solid was given the name *"sal ammoniacum"* (sal ammoniac). It is possible that this compound may have got its name from the words *"sal armeniacum"*, but it has also been suggested that it may have been named after the Egyptian god *Amon Re*. Sal ammoniac, however it got its name, was well-known to mediaeval alchemists who used it in many ways. It was obtained when organic waste was allowed to rot in wooden vats, and large, symmetrical crystals of ammonium chloride were often formed because of the presence of pectin in the wood. Sal ammoniac was essential not only for the preparation of aqua regia, which can even dissolve gold, but also in the preparation of "aurum musivum", which we know today as the golden tin disulfide SnS_2. Its ready sublimation made its use necessary for soldering metals. This chemical thus belonged in every alchemist's laboratory.

Safety Measures

Experiments in which smoke is generated should be carried out in the open air if possible; otherwise a well-ventilated fume hood should be used.

Aniline and benzoyl peroxide are extremely toxic. Other highly poisonous compounds can be formed when they react or when dyes are subjected to decomposition.

Apparatus

Fire-resistant support, test tubes with stand and fixing device, safety glasses, protective gloves.

Chemicals

(A) Charcoal, potassium nitrate, ammonium chloride
(B) Benzoyl peroxide, aniline
(C) Lactose, potassium nitrate, diatomaceous earth, indigo, auramine hydrochloride, *p*-nitroaniline red (para red).

Experimental Procedure

(A): One of the following mixtures is prepared and ignited on a fire-resistant support: either 5 g of charcoal with 3 g of KNO_3 and 2 g of NH_4Cl or 3 g of lactose with 3 g of KNO_3 and 4 g of NH_4Cl. The combustion of carbon or sugar in the presence of the oxidising agent KNO_3 generates enough heat to decompose ammonium chloride (this requires a temperature of at least 335 °C); on cooling a fine cloud of ammonium chloride falls on to the support.

(B): Two test tubes are fixed in position and filled to a height of ca. 1 cm with benzoyl peroxide; careful addition of 2 drops of aniline leads to the immediate formation of white mushroom clouds.

(C): 5 g of lactose are mixed thoroughly with 5 g of KNO_3; 2 g of diatomaceous earth are added to the mixture. The color of the smoke evolved on ignition can be varied as follows: the incorporation of 4 g of finely powdered indigo into the mixture gives a cloud of magnificent blue smoke, while green smoke is evolved when 2 g of indigo are mixed with an equal amount of the yellow auramine hydrochloride. Red smoke is formed when 10 g of *p*-nitroaniline red are used.

Explanation

The oxidising agents KNO_3 or $KClO_3$ and the fuels sugar or charcoal generate the temperature necessary for the formation of smoke; the dyes evaporate or sublime without themselves burning.

The mixtures used in fireworks for smoke generation usually consist of the following mixtures: 20–35 % by weight of lactose, 22–33 % by weight of $KClO_3$, 30–50 % by weight of the appropriate dye, 2–4 % by weight of paraffin oil, 3–10 % by weight of $NaHCO_3$ and as catalysts small amounts of MnO_2 or Fe_3O_4 (We shall not describe any smoke mixtures containing chlorinated hydrocarbons because of their detrimental effect on the environment).[1]

Reference

1 K. Menke, *Chemie in unserer Zeit*, **1978**, *12*, 13.

23

The Thermite Process

There is a theory claiming that whenever someone finds out precisely why the universe exists, and for what purpose, it will disappear on the spot and be replaced by something more bizarre and inconceivable.

There is another theory according to which that has already occurred.

Douglas Adams

Measured against that which we desire, our knowledge and understanding are clearly limited, and if in this context we wish to regard ourselves as children, then we know as well that we are growing.

Justus Liebig

I dont want to make the wrong mistake.

Yogi Berra

In the year 1827 *Wöhler* wrote as follows "On Aluminium" in the Annalen für Physik und Chemie:

The reduced mass is generally entirely molten and dark gray. The thoroughly cooled crucible is cast into a large glass full of water, in which the salt mass dissolves accompanied by the gentle evolution of a foul-smelling hydrogen gas, in the course of which a grey powder settles out; closer observation, especially in sunlight, shows this to consist apparently of nothing but tiny flecks of metal. Once it has settled, the liquid is decanted and the precipitate is transferred to a filter, rinsed with cold water, and dried. It is aluminum.

18 years later, in 1845, *Wöhler* was successful in improving his preparative method, so that instead of a grey powder he obtained beads of aluminum as large as a pinhead.[1]

On February 6th 1854 the French chemist *Henri Sainte-Claire Deville* reported to the Acédémie des Sciences in Paris on his experiments:

It is well known that Wöhler *obtained aluminum in the form of a powder when he treated the chloride with potassium; if one modifies* Wöhlers *procedure in an appropriate way the decomposition of chloride of aluminum can be regulated in such a way that there results a white-heat sufficient to cause the metal particles to fuse into spheres. Heating to a lively red-heat in a porcelain crucible the resulting mass, which consists of the metal and chloride of sodium, causes excess chloride of aluminum to escape, leaving behind a salt mass that produces an acidic reaction, and in which can be found more or less large spheres of entirely pure aluminum.*

Wöhler was very annoyed by this publication, and on March 13, 1854, he wrote in a letter to *Otto Linné Erdmann*, editor of the *Journal für praktische Chemie* (in which important articles from *Comptes rendus* regularly appeared):

You have undoubtedly already read the information in Comptes rendus *from* Deville *concerning aluminum, and noticed that it was actually new only to the French, apart from the information, not discussed more fully, that* Deville *has reduced aluminum with metals other than potassium. Under the assumption that you will include* Deville's *report in your journal, I hope I may count on your sense of justice and patriotism, so that you will add the fact that all of this (with the exception of the new reduction method) has already been known in Germany since 1845, and can be found in vol. 53 of the* Annal. d. Chem. u. Pharm. *Do me the favor on this occasion of reading my brief note. I have heard that even in the ordinary newspapers much noise is being made about* Deville's *new discovery. As soon as he succeeds in preparing aluminum by the pound or the hundredweight no one more than I will acknowledge this as a very important discovery, and no one will be any more pleased. But I fear that we are still far removed from that point, and will remain far from it so long as one is required to use as a starting material only a substance as difficult to prepare in quantity as chloride of aluminum.*

In no. 8 of the Comptes rendus *(février) you will also have read a commentary from a Mr.* Schratz *au nom de son oncle de* Völher *[sic!] and associated remarks by* Dumas. *What that signifies is completely incomprehensible to me. I know no one by this name, and I am myself the uncle to no one apart from a young boy who can scarcely walk. I think that a little bit of French deviltry lies behind this. Moreover, I have written to* Dumas *that I had no part in this commentary and I am not the uncle of Mr.* Schratz.

Liebig, too, encouraged his friend to take action against the French aluminum spectacle, and to set the story of the discovery of aluminum straight. *Wöhler* responded to him as follows:

To write something about this matter for the Annalen *as well, that is something to which I cannot really commit myself. These commentaries are always more or less petty, and there is also nothing one could say without applying a few swift kicks to this French swagger and vanity, thereby causing offense to this sensitive nation. I already regret having written the letter to* Dumas; *my main intention was to let him know that the forged commentary by Mr.* Schratz *did not stem from me. That which needs to be said has actually already been said by* Erdmann *in his journal.*

Wöhler had written to *Dumas*:

I have been following with the liveliest of interest the observations of Mr. Deville *regarding aluminum. It is quite remarkable that this chemist has struck medals from this metal [i.e.., from aluminum]. Since I assume that you [i.e., Dumas] are in possession of several samples of his [i.e., Deville's] aluminum whether in rolled form or as wire will risk saying to you here that you would give me great pleasure if you would send me a tiny piece of this metal so that I could show it in my lectures as a precious memento of* Dumas.

However, the letter was reproduced in somewhat modified form in the French journal:

I have been following with the liveliest of interest the aluminum researches of Mr. Sainte-Claire Deville. *I was greatly astonished to learn that he succeeded in striking a medal from pure aluminum. I cannot imagine that he [i.e.,* Deville*] has managed to prepare a metal that was known previously only as a dusty powder with no metallic luster in such a form that it possesses a high degree of rigidity, and is at the same time ductile and malleable. If you [i.e.,* Dumas*] could help me acquire a small sheet of aluminum it would give me the greatest pleasure, and I would proudly show it to my students during the lecture.*

The alteraction ended when *Wöhler* received an aluminum medal as a gift from *Deville*. On the front of the medal there was a portrait of the French Emperor *Napoleon III*, and the reverse side was engraved with the inscription "Wöhler 1827". This is reported in two letters from *Wöhler* to *Liebig*:

My essay for the Annalen ... *has actually become irrelevant, despite which I will allow it to be printed, albeit with a few alterations and sweeteners for* Deville. *Because the latter, at the instigation of* Dumas, *had the great courtesy to present me with a medal struck from aluminum. It is the size of a two thaler coin and somewhat thicker; inscribed on the obverse side with the head of* Napoleon III, *on the reverse with my name and the year 1827. I admit that I was astonished in the highest degree to see such a mass of aluminum. Its lightness leads one to think that it must be hollow or made from silvered Paquèn. The color, incidentally, is definitely not silvery, but bluish, rather like that of Britannia metal. I could not avoid mentioning this gift in a note to the essay, in order from my own perspective to confirm that* Deville *has truly succeeded in preparing such a large mass of aluminum. I have also thanked him in a manner signifying my acknowledgment of the credit due to him, and in such a way that he will probably cause the letter to be printed as a means of enhancing his reputation, which I would have nothing against, since one is obliged to praise that which is praiseworthy.*

I received from Kopp *today the latest issue of the* Annalen *containing the note from me about aluminum. It reminded me that I had recently forgotten to tell you about the further course of the aluminum-intrigue, and to explain to you why I let that note be published and not the longer discussion I sent you in manuscript form. The main reason is that I don't want to appear hostile with respect to* Dumas, *to offend him or expose him, irrespective of how much he has apparently earned it. Because he appears to be the chief sinner in this matter, who wants to confer the whole aluminum discovery on the French, or at least on his pupil* Deville, *and who may at the same time have other personal objectives in mind. The fact that they arranged to have the Emperors picture impressed upon the medal was certainly a subject of careful consideration, and the value that he [i.e.,* Dumas] *assigns to the work of* Deville *is apparent from the comments in a letter ... in which he characterizes the discovery of aluminum as one of the greatest of the century, and equates it with telegraphy, electrometallurgy, etc. He pays me great compliments in the process, and designates me as le père de l'aluminium, while at the same time in Paris he attempts to cast suspicion on this fatherhood in that he communicates my postscript in such a twisted and distorted way, as you well know. Given the high regard that I have for* Dumas *as a chemist, I therefore believed it better under such circumstances to let the matter drop, and thereby ignore his involvement. I confess to you also, that another consideration was the fact that you enjoy a friendly relationship with* Dumas, *and that such an open collision between him and me would certainly not have been agreeable. The note that has now appeared in the* Annalen *has as its sole purpose to grant some public recognition to* Deville *for the courtesy with which he sent me the medal, on the reverse side of which he had my name engraved along with the year 1827. That is to say, they have protected themselves very well by letting this medal be struck.*

Safety Precautions

Safety glasses and leather gloves should be worn. Accidents have often happened when this experiment was being carried out. The chemicals must be absolutely dry, otherwise the mixture can be catapulted out of the crucible and cause third degree burns! This experiment should on no account be carried out by schoolchildren or students, who must keep at least 3 m away from it.

Apparatus

Clay crucible (90 mm in diameter at the top, height 120 mm, with a hole 6–8 mm in diameter in its base), stand, boss, ring clamp for the crucible, iron sandbox(base 80 × 80 cm, height 15 cm), crucible tongs, mortar and pestle, gas burner, hammer, magnet, safety glasses, leather gloves.

Chemicals

Aluminum shot, aluminum powder, Fe_2O_3, sparkler, 20 kg sand, magnesium ribbon (about 10 cm in length).

Experimental procedure[2]

The clay crucible is attached to the stand by means of the ring clamp. The stand is placed in the sandbox, which is filled with sand to a height of about 5 cm. The mixture of 20 g of dry iron oxide and 6 g of aluminum shot (which should if necessary be dried *separately* at about 150 °C) is ground in the mortar and introduced into the clay crucible, the hole in the base of which has previously been covered by a piece of filter paper. 5 g of aluminum powder are then poured over the mixture and one end of the magnesium ribbon pushed into the mixture. The other end projects above the upper edge of the crucible and is ignited with a gas burner. The reaction is extremely violent and is accompanied by the formation of a large amount of sparks. The liquid mass which is formed drops through the hole in the crucible on to the sand and forms a glowing regulus.

   After cooling in the air the regulus is held under water using the tongs and shattered with the hammer on a support so that the slag and iron can be separated. The iron can be identified with the help of a magnet.

   When this experiment is to be carried out at school we suggest the use of a thermite mixture. 30 g of this mixture are used instead of the mixture of iron oxide, aluminum shot and aluminum powder. The thermite mixture is ignited by carefully throwing in a lighted sparkler.

Explanation

The reaction takes place according to the following equation:

$$Fe_2O_3 + 2\ Al \rightarrow Al_2O_3 + 2\ Fe \qquad \Delta H = -730\ kJ/mol$$

References

1 W. Schäfke, T. Schleper, M. Tauch, *Aluminium,* Köln, **1991**.
2 F. Brandstätter, L. Sternhagen, *Chemische Schulversuche,* Wien, **1950**.

24

Antimony Triiodide

In proceeding to consider the metals, which in the inorganic world have the almost exclusive prerogative of appearing colored, we find that, in their pure, independent, natural state, they are already distinguished from the pure earths by a tendency to some one color or other.

Generally, however, these appearances of color are of so mutable a nature that chemists look upon them as deceptive tests, at least in the nicer gradations. For ourselves, as we can only treat of these matters in a general way, we merely observe that the appearances of color in metals may be classed according to their origin, manifold appearance, and cessation, as various results of oxydation, hyper-oxydation, ab-oxydation, and de-oxydation.

Johann Wolfgang von Goethe, "Theory of Colors" (1810)

The alchemist's laboratory at the court of the Duke *Friedrich I* of Württemberg in the early 17th century included "geschmelzt" antimony and "Bergantimon" among its chemicals. The latter was probably diantimony trisulfide Sb_2S_3, which was known as antimonite or grey antimony and had been introduced into medicine by *Paracelsus*.[1] "Antimony butter", antimony trichloride, was used medicinally as a caustic agent. The preparation of the related antimony triiodide SbI_3 presents an interesting play of colors.

Apparatus

Large test tube with holder, mortar and pestle, spatula, 100-mL round-bottomed flask with reflux condenser, 100-mL beaker, funnel with paper filter, suction flask with glass frit, gas burner, ice bath, safety glasses, protective gloves.

Chemicals

Antimony metal in small pieces, iodine flakes, toluene, diethyl ether.

Experimental Procedure

A pinch of antimony and of iodine are mixed, powdered, and heated carefully in the test tube. During the violent reaction violet iodine vapor is formed; this rapidly condenses and the inner wall of the glass tube is rapidly coated with a deep red layer of antimony triiodide SbI_3.

A pure product is obtained in almost quantitative yield according to the following procedure: 3.5 g of iodine are heated with 1.0 g of finely powdered antimony in 40 mL of toluene at reflux until the solution has a deep orange color. The mixture is filtered hot through a filter paper and the filtrate cooled in the ice bath. After a short time beautiful scarlet crystals of antimony triiodide separate; these are filtered off using a glass frit and washed with a little ice cold diethyl ether. If care is taken the yield can approach the theoretical maximum of 4.1 g. The same amount of SbI_3 is always obtained, even if the amount of iodine is varied (for example 4.0 g or 5.0 g).

Explanation

Under the conditions of the experiment antimony reacts as follows:

$$2 \; Sb + 3 \; I_2 \rightarrow 2 \; SbI_3$$

The almost invariant yield when varying amounts of iodine are used provides a clear demonstration of the Law of Constant Proportions (*J. L. Proust*, 1799).

Waste Disposal

The pure end product can be stored in the collection of laboratory chemicals (the SbI_3 in the test tube can also be isolated by treating it with ice-cold toluene by drying); toluene is purified by distillation or transferred to the container used for storing organic solutions.

Reference

1 H.-G. Hofacker, *"..... sonderlich hohe Künste und vortreffliche Geheimnis", Alchemie am Hof Herzog Friedrichs I. von Württemberg (1593–1608)*, Wegrahistorik-Verlag, Stuttgart, **1993**.

25

The Many Colors of Vanadium

A sonnet written by the French poet *J. N. A. Rimbaud* (1854–1891[1]) shows how images and associations arising from the use of color symbolism can heighten the reader's pleasure:

<table>
<tr><td>Voyelles</td><td>Vowels</td></tr>
</table>

A noir, E blanc, I rouge, U vert, O bleu, voyelles,   *A black, E white, I red, U green, O blue;*
Je dirai quelque jour vos naissances latentes: *Someday I'll tell your latent birth O vowels;*
A, noir corset velu des mouches éclatantes   *A, a black corset hairy with gaudy flies*
Qui bombinent autour des puanteurs cruelles,   *That bumble round all stinking putrefactions,*
Golfes d'ombre: E, candeurs des vapeurs et des tentes,   *Gulfs of darkness; E, candors of steam and tents*
Lances des glaciers fiers, rois blancs,   *Icicles proud spears, white kings,*
frissons d'ombelles;   *and flutter of parasols;*
I, pourpres, sang craché, rire des lèvres belles   *I, purple blood coughed up, laughter of lovely lip*
Dans la colière ou les ivresses pénitentes.   *In anger or ecstatic penitence;*
U, cycles, divine vibrations of virescent seas,
Peace of the pastures sown with animals, peace
Of the wrinkles that alchemy stamps
on studious brows;
O, Clarion supreme, full of strange stridences,
Silences crossed by Angels and Worlds:
Omega, the violet ray of His Eyes!

Many, many years ago there lived in the far North the beautiful and greatly beloved goddess *Vanadis,* a member of the race of gods from Nordic mythology called the Vanes. The metal vanadium, which is now named after her, was initially called *"panchromium"* by its Mexican discoverer *A. M. del Rio* (1801) because of the many colors the salts exhibited; the exact proof that it was in fact a previously unknown element had to wait until 1830/31 and is due to the Swedish chemists *J. J. Berzelius* (1779–1848) and *N. G. Sefström* (1787–1845), one of his pupils.

Apparatus

250-mL glass measuring cylinder, glass rods, safety glasses, protective gloves.

Chemicals

Ammonium metavanadate, zinc granules, concentrated sulfuric acid, half-concentrated hydrochloric acid, distilled water.

Experimental Procedure

1 g of ammonium metavanadate, NH_4VO_3, is suspended in 200 mL of water. The initially colorless solution turns yellow when 2 mL concentrated sulfuric acid are added, a small amount of a red precipitate being formed. The addition of 6–8 zinc granules followed by first 25 mL and then a further 40–50 mL of half-concentrated hydrochloric acid leads to a reaction which is accompanied by violent gas evolution. The mixture first turns sky-blue, then green and finally turquoise for a short time a pale lilac color is observed, but this is rapidly followed by a return to turquoise (colored figure 9).

Explanation

The initial yellow coloration of the solution is due to the presence of the dioxo-vanadium(V) ion, VO_2^+; the addition of H_2SO_4 to the NH_4VO_3 solution causes a small amount of red V_2O_5 to be precipitated. The VO_2^+ ions are initially reduced by the nascent hydrogen into the sky-blue hydrated oxovanadium(IV) ion, $[VO(H_2O)_5]^{2+}$ and then to the green $[V_{aq}]^{3+}$ ion via the oxidation states $+V$, $+IV$ and $+III$. The lilac coloration is due to the presence of the $[V_{aq}]^{2+}$ ion, which has only a short lifetime under the experimental conditions. The Zn^{2+} ions formed in the oxidation of metallic zinc are colorless and thus have no effect on the color of the solution. Since the various steps in the reduction of the vanadate are not separated, the colors are not quite the same as those observed when the solutions of the various vanadium salts are prepared in the absence of air.[2]

Waste Disposal

The solutions are rendered slightly alkaline by adding a solution of sodium carbonate. They are decanted, the precipitates transferred to the container used for collecting less toxic inorganic waste, and the mother liquor poured down the drain.

Reference

1 Cited by W. Drost, *Diagonal*, Journal of the Universität-GH Siegen, **1993**, *1*, 81.
2 F. A. Cotton, G. Wilkinson, *Advanced Inorganic Chemistry*, John Wiley & Sons, New York, Chichester, Brisbane, Toronto, Singapore **1988**, p. 667 et seq.

26

The Stepwise Reduction
of Potassium Permanganate
in an Alkaline Medium

How should one speak of colors? It is in a sense reasonable that only the blind should engage in arguments about this, just as the rest of us debate metaphysics. But those who can see are utterly unaware that the word and what they actually see cannot be measured against each other.

The poet *Paul Valery* is reflecting on the works of the painter *Edouard Manet*;[1] he will certainly not have known the problems which the chemist has when trying to observe and interpret the variations in color shown by the compounds derived from the element manganese, which was discovered in 1774 by *J. G. Gahn*. However, "lapis manganensis" (manganese dioxide) was certainly well known to the glassmakers of antiquity, mainly because of its compensatory (decolorising) properties.

Apparatus

Two glass cylinders 30 × 3.5 cm, two beakers (600-mL and 50-mL), magnetic stirrer, 25-mL burette, glass rods, safety glasses, protective gloves.

Chemicals

6 M NaOH (120 g in 500 g solution), 0.1 % H_2O_2 solution (1 mL of 30 % H_2O_2 in 350 mL of water), 6 mol/L acetic acid (36 mL glacial acetic acid diluted to 100 mL with water), 0.05 g $KMnO_4$ (dissolved in 1 mL of water), ice.

Experimental Procedure

The sodium hydroxide solution is cooled to 0 °C and the potassium permanganate is added; the resulting solution is purple. About 1 mL of the hydrogen peroxide solution is added from the burette with constant stirring until the mixture turns green. This solution is used to fill one of the glass cylinders half full; 20 mL of acetic acid are carefully added so that two layers are formed. If the upper phase is stirred slowly with the glass rod its color changes to the reddish blue typical of $KMnO_4$ and is clearly different from the dark green of the lower phase.

If a further 2 mL of H_2O_2 are now added to the alkaline solution in the beaker, a further color change to blue occurs. This solution is used to fill the second glass cylinder half full. Addition of 20 mL of acetic acid as before, followed by careful stirring, produces a green upper phase, and if this is again treated with 20 mL of acetic acid and stirred (see above) then three differently colored layers can be seen (at best against a white background): the lower blue layer, then the pale green phase and finally the violet layer which has the color of the original potassium permanganate (colored figure 10).

Explanation

In strongly alkaline solution the MnO_4^- ion is first reduced by hydrogen peroxide to the green manganate(VI) ion MnO_4^{2-}. The addition of acetic acid causes disproportionation to give permanganate and the yellowish-brown manganese(IV) oxide (MnO_2), the color of which is visible at the phase boundary. The addition of further H_2O_2 leads to the blue, relatively unstable MnO_4^{3-} ion, which is reconverted to the manganate(VI) ion under acid conditions: the addition of further acid gives the violet potassium permanganate as described above. The various layers – clearly separated phases are not formed – demonstrate the various oxidation states of the MnO_4^{x-} ions reduced by hydrogen peroxide + VII for x = 1, + VI for x = 2, + V for x = 3. The stable final state for all these processes is manganese(IV) oxide, which is formed in pale yellowish-brown colloidal state. The indicated cooling is essential if this experiment is to succeed well. In acidic medium $KMnO_4$ is readily reduced to Mn^{2+} ion, a circumstance that can be taken advantage of in oxidimetric titration.

Disposal

Neutralize with dilute sulfuric acid and rinse down the drain.

Reference

1 Cited according to M. Dobbe, *Diagonal,* Journal of the Universität GH Siegen, **1992**, *2*, 75.

27

A Reversible
Blue-and-Gold Reaction

Vincent van Gogh writes as follows to his brother *Theo* on August 11th 1888:

> *I want to make a picture of a friend, an artist who dreams of great things, who works the way a nightingale sings because that is his nature. The man will be blonde. I want to instill in the picture the wonder and the love that I feel for him.*
>
> *So I will paint him first just as he is, as faithfully as I can. But that will not be the end of the picture. In order to complete it I will now become an independent colorist.*
>
> *I exaggerate the blonde of the hair, come to orange tones, to chromes, to a pale lemon-yellow.*
>
> *Behind the head, instead of the mute wall of the shabby room I paint the infinite, I make a simple background of the richest, most penetrating blue I can create, and through this simple combination the blonde, luminous head against the rich blue background acquires something of the mysterious, like the stars in a deep blue heaven.*
>
> *I have proceeded in a similar way on the peasant picture. Admittedly without here wanting to summon the mysterious effect of a pale star surrounded by the infinite. But I have imagined the dreadful man I had to picture in the red-hot harvest season, in the full heat of midday.*
>
> *That is the reason for the blazing orange like red-hot iron, for the tones of luminescent antique gold in the dark.*
>
> *Oh, my dear brother and the good people will see nothing but caricature in this exaggeration ...*

The chemist cannot of course come anywhere near to creating such a wonderful composition in blue and gold as produced by nature or by this great artist, but he can remind us of this imposing play of colors with the help of a simple redox reaction.

Apparatus

1-L beaker, heater-stirrer hotplate with stir bar (oder gas burner as heat source and stirring rod), thermometer, 1-mL pipette, safety glasses, protective gloves.

Chemicals

1 mol/L $KNaC_4H_4O_6 \cdot 4 H_2O$ (282 g in 1 L aqueous solution), 1 mol/L $CuSO_4 \cdot 5 H_2O$ (250 g in 1 L aqueous solution), 3 % hydrogen peroxide solution.

Experimental Procedure

60 mL of the aqueous sodium potassium tartrate solution and 40 mL of H_2O_2 are mixed in the beaker. The mixture is heated to 50 °C and 1 mL of copper sulfate solution added with constant stirring. The solution turns sky blue, warms to 80 °C, foams because of the evolution of gas, and suddenly turns to a cloudy golden orange color. If a further 40 mL of hydrogen peroxide solution are added the orange precipitate dissolves, the blue color reappears and shortly afterwards the solution again turns golden orange.

This cyclic process can be repeated 5 to 6 times if the reaction solution remains sufficiently hot (at least 70 °C) (colored figure 11).

Explanation

The sodium potassium tartrate initially forms the soluble blue copper(II) tartrate complex with the copper sulfate; the color of this complex is a deeper blue than that of the hydrated Cu^{2+} ion. At the temperature of the experiment reduction to the beautiful golden orange Cu_2O occurs, and at the same time oxygen and (rather less) CO_2 are formed in the redox process; these are responsible for the foaming of the solution. This reversible color reaction can be repeated several times when further hydrogen peroxide is added. The pH of the system increases from 5 to about 9.[1]

Waste Disposal

The reaction mixture is left to stand for a considerable time; the solution is decanted from the precipitate and the latter transferred to the container used for collecting less toxic inorganic substances. The clear solution is poured down the drain.

Reference

1 M. C. Sherman, D. Weil, *J. Chem. Educ.,* **1991,** *68,* 1037.

28

The Two-Color Formaldehyde Clock

Such a play of colors and lights, different seasons, different hours of the day – the lines of the far horizon where the faint-tinged edge of the landscape loses itself in the sky. As I slowly hobble up the lane toward day-close, an incomparable sunset shooting in molten sapphire and gold, shaft after shaft, through the ranks of the long-leaved corn, between me and the west.

Another day. – The rich dark green of the tulip-trees and the oaks, the grey of the swamp-willows, the dull hues of the sycamores and black-walnuts, the emerald of the cedars (after rain) and the light yellow of the beeches.

Walt Whitman (1819–1892), "Specimen Days" (1882)

Safety Measure

Formaldehyde is toxic. The experiment should be carried out in well-ventilated hood.

Apparatus

Two 500-mL beakers, four 250-mL graduated beakers, dropping pipettes, stock bottles for indicator solutions, two 1-L measuring cyclinders, safety glasses, protective gloves.

Chemicals

Formalin (formaldehyde solution, 37 %), sodium sulfite, sodium hydrogen sulfite, ethylenediamine tetraacetate (EDTA), thymolphthalein solution, *p*-nitrophenol, phenolphthalein, ethanol.

Solution A: 22 mL of formaldehyde solution are diluted to 1 liter with water and allowed to stand for 24 h to allow complete depolymerisation to occur. The HCHO concentration is 0.3 mol/L.

Solution B: 20.8 g of sodium hydrogen sulfite and 6.3 g of sodium sulfite are dissolved in water to give one liter of the solution; this is 0.2 mol/L with respect to $NaHSO_3$ and 0.05 mol/L with respect to Na_2SO_3. 0.37 g of EDTA are added. The solution can be kept for about 3 days.

Indicator Solution I: 10 mL of thymolphthalein solution (1.5 g in 100 mL of ethanol) are mixed with 45 mL of *p*-nitrophenol solution (5.0 g in 100 mL of ethanol). The green color can be varied slightly by changing the composition of the mixture.

Indicator Solution II: 30 mL of phenolphthalein solution (1.5 g in 100 mL of ethanol) are stirred with 20 mL of *p*-nitrophenol solution; here again the red color can be varied slightly if required.

Experimental Procedure

Variation I: 20 drops of the indicator I are added to 200 mL of solution A. This solution is mixed with a further 200 mL of solution B, and the resulting mixture immediately turns golden yellow. After 30–40 seconds the color changes to green.

Variation II: 20 drops of the indicator II are added to 200 mL of solution A. A further 200 mL of solution B are added, and as above a golden yellow solution is obtained. After 30–40 seconds the color changes to red.[1]

The colors appear more striking if a direct light is shone on the reaction vessels which are held against a dark background.

Waste Disposal

The solutions are neutralised with dilute acid and poured down the drain.

Reference

1 J. J. Fortman, J. A. Schreier, *J. Chem. Educ.*, **1991**, *68*, 324.

The Colors Black, Red, and Gold

On March 9th 1848, under pressure from the bourgeois revolution which was just starting, the German Bundestag proclaimed the German federal colors to be black, red and gold. Thus the colors of the flag of the Jena student society, which had been adopted by the Lützow Free Corps for its uniforms, were chosen as the symbol of the national and liberal political movement in the country. The German National Assembly, which was constituted in the Paulskirche in Frankfurt on May 18th of that year of revolution as the result of general, free and direct elections, confirmed "black-red-gold" as the German national colors. Beneath this flag the Reichsverweser *Archduke Johann*, acting on behalf of the Assembly of the Reich, promulgated the law *"The Basic Rights of the German People"*[1] which marked the beginning of a democratic Germany. According to this tradition the colors "black-red-gold" embodied the highest symbol of the state during the Weimar Republic, and since 1949 the black-red-gold flag is the national flag of the Federal Republic of Germany.

Safety Measures

Arsenic and its compounds are highly toxic.

Apparatus

Three 1-L beakers, six 150-mL beakers, three 50-mL beakers, safety glasses, protective gloves.

Chemicals

Potassium iodate, sodium sulfite, starch, concentrated sulfuric acid, glacial acid, sodium arsenite, sodium thiosulfate. The indicator solution II and the solutions A and B from the previous experiment are also required.

Experimental Procedure

200 mL of water are placed in one of the large beakers, followed by 20 mL of a 1 % starch solution. 100 mL of a solution of sodium sulfite in sulfuric acid (made by adding 1.6 g Na_2SO_3 (anhydrous) to a mixture of 1 L water, 10 mL ethanol and 4 g concentrated H_2SO_4) are added with stirring, and finally the

X   Die Grundrechte des Deutschen Volkes. Lithographie von Adolf Schroedter, Mainz 1848

The Basic Rights of the German People

solution of 0.5 g KIO$_3$ in 100 mL water is added. This mixture turns almost **black**.

200 mL of the solution A are placed in the second large beaker; 25 drops of the indicator solution II (see previous Experiment 28) are added, followed by slow addition of 200 mL of solution B. The mixture is stirred for a few seconds, and after about a minute the **red** color is formed.

200 mL of water are now added to the third large beaker, followed by 30 mL of glacial acetic acid. Solutions of sodium arsenite (10 %) and sodium thiosulfate (saturated, 100 mL in each case) are now poured in. The mixture is stirred for a short time and turns **golden yellow** after about 45 seconds.

Explanation

The (deep blue-)black solution in the first beaker is due to the starch-iodine complex formed in the reaction; the red color formed in the second beaker is caused by the indicator phenolphthalein and the *p*-nitrophenol component in an alkaline medium, while elemental sulfur and As$_2$S$_3$ are mainly responsible for the golden yellow color generated in the third beaker.

Waste Disposal

The combined mixtures from the three beakers are boiled for about 10 minutes with milk of lime and 30 mL of 30 % hydrogen peroxide. After cooling the precipitate is separated and transferred to the container used for collecting toxic inorganic waste. The solution is poured down the drain.

Reference

1 Exhibition catalog *"Fragen an die deutsche Geschichte"*, **Plate X,** Bonn **1990.**

30

Reactions with Iodine

In 1811 the Parisian kelp burner *B. Courtois* distilled off a deep violet colored vapor from a mixture of seaweed ash, soda and concentrated sulfuric acid; on cooling the vapor solidified to give metal-like crystals. This was the discovery of iodine. Two years later *J. Gay-Lussac* suggested that this new substance belonged to the halogen series and gave it the name iodine (from the Greek *ioei-des*, violet-colored). As it had been known for a long time that seaweed provided an extremely useful remedy for treating goiter, potassium iodide was introduced as a drug as early as 1819). The iodine-containing thyroxine is present in the thyroid hormone; (its structure see below) was determined by *C. R. Harington* in 1926. If the supply of iodine is insufficient the thyroid gland increases in size. The addition of small amounts of potassium iodide to table salt is thus a useful prophylactic measure.

The intense color of iodine and its color changes in solution led rapidly to applications in chemical analysis. Several such reactions will be demonstrated in this chapter.

Safety Measures

Toluene and carbon disulfide are extremely toxic. The experiments should be carried out in a well – ventilated hood.

Apparatus

Two 250-mL Erlenmeyer flasks with ground-glass stoppers, ten large test tubes with cork stoppers, seven small test tubes, two test tube stands, porcelain spatula or spoon, safety glasses, protective gloves.

Chemicals

Potassium iodide, crystalline iodine, sodium thiosulfate, chlorine and bromine water prepared beforehand and kept in small bottles, starch solution, distilled water, *n*-hexane or *n*-heptane, carbon disulfide, cyclohexene, toluene, acetone, diethyl ether.

The structural formula of thyroxine, the amino acid present in the thyroid gland.

Experimental Procedure

Reaction Sequence A: 100 mL of a 10 % aqueous solution of KI are placed in each Erlenmeyer flask. 100 mL of chlorine water are added to one flask, 100 mL of bromine water to the other; in both cases the solution immediately turns brown. The contents of the two flasks are now transferred to the five large test tubes, to which the following substances are added: test tube 1, freshly pre-pared starch solution; test tube 2, a few grams of solid KI; test tube 3, diethyl ether; test tube 4, toluene; test tube 5, *n*-hexane (or *n*-heptane). The contents of the first test tube turn to a deep blue-black, while the brown color in the second becomes much deeper. In the other test tubes two layers are formed; in each case the lower aqueous layer is much lighter in color, while the upper organic layers are colored yellow(test tube 3), wine-red (test tube 4) and violet (test tube 5) respectively.

Reaction Sequence B: A few crystals of iodine are dissolved in each of the seven smaller test tubes, which contain different solvents.

Test tube 1 contains *n*-hexane (or *n*-heptane), test tube 2 CS_2; in both cases a deep violet color is formed. Test tubes 3 and 4 are filled with toluene and cyclo-hexene respectively: iodine colors these solvents wine-red. Test tube 5 contains acetone and test tube 6 diethyl ether: here the iodine solutions are yellow.

In test tube 7 the reaction of a concentrated KI solution with solid iodine leads to a deep brown coloration; this solution can be decolorised by adding an excess of $Na_2S_2O_3$.

Explanation

Elemental bromine and chlorine set iodine free from aqueous iodide solutions due to their higher redox potential:

$$(1)\ X_2 + 2\ I^- \rightarrow I_2 + 2\ X^-$$

Free iodine reacts with starch solution to give the blue-black amylose-iodine complex, which contains iodine in the form of an I_5^- chain.[1] The amylopectin also present in starch reacts with iodine to give a brown-violet color. In concen-trated aqueous alkaline iodide solutions and in alcoholic solution iodine has an intense brown color; at high dilutions this color changes to yellow because of the formation of I_3^- ions. Solutions of iodine in acetone and pyridine are also yellow because of partial cleavage of I-I bonds. Toluene and cyclohexene are good π-donors and thus form charge transfer complexes with iodine: these are wine-red. In non-polar solvents such as alkanes, carbon tetrachloride or CS_2 the color of the iodine molecule is visible. All this was recognised by *E. Beckmann* as early as 1890.[2]

Thiosulfate solutions decolorise iodine, the thiosulfate ion being converted to tetrathionate:

$$(2)\ 2[S_2O_3]^{2-} + I_2 \rightarrow [S_4O_6]^{2-} + 2\ I^-$$

Waste Disposal

The organic phases are transferred to the container used for collecting halogenated solvents and the aqueous layers poured down the drain.

References

1 R. C. Teitelbaum, S. I. Ruby, T. J. Marks, *J. Am. Chem. Soc.*, **1978**, *100*, 3215.
2 E. Beckmann, *Z. Phys. Chem.*, **1890**, *5*, 76.

Methylene Blue: A Dyestuff that Made Medical History

The young country doctor *Robert Koch* discovered the tubercle bacillus with the help of methylene blue. He demonstrated that the smallest forms of life are the source of all infectious diseases. One of his pupils was *Paul Ehrlich*, who one day noticed that methylene blue had a surprising affinity for living cells, which it colored an intense blue. "If only certain cells are colored", thought *Ehrlich*, "then may there not be dyestuffs which color only the carriers of illnesses and at the same time destroy them without attacking the body's own cells?"

The revolutionary idea led to the development of chemotherapy, one of the greatest advances in medical science.

Methylene blue was first synthesised by the German chemist *Heinrich Caro* in 1876, and *Ehrlich* could obtain as much dye as he required from the Farbwerke Hoechst, who had been manufacturing it since 1885. *Ehrlich* eventually developed the famous Salvarsan®, the most effective anti-syphilis drug. This he did in 1910, thus becoming the *father of chemotherapy*.[1] However, methylene blue is not only a good coloring agent for living cells but also an excellent redox indicator in aqueous solution.

Apparatus

2-L round-bottomed flask with rubber stopper, cork ring, thermometer, tripod with wire gauze, stand, stopwatch, safety glasses, protective gloves.

Chemicals

Glucose, NaOH pellets, distilled water, 0.2 % aqueous solution of methylene blue.

Experimental Procedure

10 g of solid NaOH are added to 750 mL of water in the round-bottomed flask, followed by 40 g of glucose. This mixture is treated with 10 mL of the methylene blue solution and the flask stoppered. If the flask is now shaken vigorously, the initially colorless solution turns deep blue but after a short time again

The redox system methylene blue-leucomethylene blue

becomes colorless. It can be shown that the time required for the color to disappear depends on the number of shaking operations carried out: this can be made clear by plotting the number of shaking operations against the decolorisation time, which is determined using a stopwatch. If the mixture is heated in steps of 10 °C to maximum of 50 °C the color change takes place much more quickly and is clearly temperature-dependent. The flask must be opened before each experiment to make sure that a sufficient amount of oxygen is present.

Explanation

The glucose reduces methylene blue to the leucobase and is itself oxidised to gluconic acid, which is converted in alkaline solution to sodium gluconate. The oxygen in the air is activated when the flask is shaken and converts the leukobase into methylene blue. The blue color remains longer when the shaking is more frequent. According to the rules of chemical kinetics a rise of 10 °C in temperature causes the reaction rate to rise by a factor of two to four (depending on the reaction type); this is shown by the much shorter time required for the decolorisation in this experiment when the temperature is raised. The reaction rates in this coupled redox process can readily be compared when a graph is prepared as described above. At temperatures above 50 °C complex oxidation processes occur which are no longer subject to a simple correlation.

Waste Disposal

The reaction mixture is neutralised and poured down the drain.

Reference

1 Contribution from the Hoechst AG, *Diagonal,* Journal of the Universität-GH Siegen, **1992**, 2, 92.

32

The Volcano Experiment

Because of the eternal flame burning inside it seems to be red.

Dante "Inferno"

The greatest volcanic eruption in history took place between Sumatra and Java on the small island of Krakatoa between August 26th and 28th 1883. At 1 p.m. on August 26th there was a huge explosion, the sound of which could be heard as far as 160 km away. An hour later an even more violent explosion occurred, spewing out ash and lava to a height of 27 km. During the whole night short but increasingly violent explosions followed; these could be heard even on Sumatra and Java. Within a radius of 160 km so much ash fell that the sun could scarcely be seen. In Djakarta it was necessary to burn torches and lamps for a whole day, and on the coast for two!

The greatest explosion took place on August 27th, throwing up ash to a height of 80 km; an area twice as big as Germany was covered with a torrential rain of ash and lava. The sound waves produced are said to have gone around the earth several times, and the noise of the explosion could be heard 4800 km away. This huge explosion was followed the same day by two more which were almost as violent. The uninhabited island of Krakatoa had been completely destroyed, and seaquakes laid waste to the heavily populated sea coasts of Java and Sumatra. More than 36,000 people drowned. A great wave caught a gunboat which lay at anchor off the town of Telukbetung and propelled it into the middle of the town; a second, larger wave followed and cast the boat into the middle of the jungle. On the north-west coast of Java the waves reached a height of 40 m, and in Pepper Bay the waves rushed 16 km inland.

Safety Precautions

This experiment should only be carried out in a well-ventilated hood, since chromates and dichromates are strongly cancerogenic. Safety glasses and protective gloves must be worn.

Apparatus

150-mL beaker, fire-resistant support, safety glasses, protective gloves.

Chemicals

Ammonium dichromate $(NH_4)_2Cr_2O_7$, acetone.

Experimental Procedure

A 150-mL beaker is filled half-full with $(NH_4)_2Cr_2O_7$. The contents of the beaker are then poured on to the fire-resistant support and formed into a cone. The tip of the cone is wetted with a few mL of acetone so that the ammonium dichromate can more readily be set light to by a gas burner.

Explanation

The ammonium dichromate decomposes to Cr_2O_3, setting free nitrogen and water. The green reaction product occupies a much larger volume than the starting material.

$$(NH_4)_2Cr_2O_7 \rightarrow N_2 + 4\ H_2O + Cr_2O_3$$
$$\text{orange} \hspace{4cm} \text{green}$$

Waste Disposal

The residue is transferred to the container used for collecting toxic inorganic waste.

33

Catalytic Decomposition of Ammonia in the Presence of Oxygen

A molecule swims, dispersing its functionality, scattering its reactive centers. Not every collision, not every punctilious trajectory by which billiard-ball complexes arrive at their calculable meeting places leads to reaction. Most encounters end in a harmless sideways swipe. An exchange of momentum, a mere deflection. And so it is for us.

Roald Hoffmann, "The Metamict State" (1987)

It is a profound and necessary truth that the deep things in science are not found because they are useful; they are found because it was possible to find them.

Robert Oppenheimer

Safety Precautions

Concentrated ammonia vapor can cause severe damage to mucous membranes and the eyes. Wear safety glasses. The experiment must be carried out in well-ventilated hood.

Apparatus

500-mL Erlenmeyer flask, platinum coil, nickel wire, stand, clamps, bosses, gas inlet tube, safety glasses, protective gloves.

Chemicals

Concentrated ammonia solution, oxygen from a gas cylinder.

Experimental Procedure

150 mL concentrated ammonia solution are placed in the Erlenmeyer flask. The platinum coil is attached to a piece of nickel wire and hung in the flask so that it is about 3 cm above the surface of the liquid. Before the spiral is intro-

duced into the flask it is heated with a gas burner. The ammonia vapor decomposes and the coil begins to glow.

If bursts of oxygen are allowed to flow into the flask through the glass tube the decomposition becomes much more violent and a jet of flame is formed. The glowing coil can be observed much better when the room is darkened.

The decomposition of ammonia during this experiment takes place predominantly according to the following equation:

$$4\ NH_3 + 3\ O_2 \rightarrow 2\ N_2 + 6\ H_2O$$

Waste Disposal

The remaining ammonia solution can be used for other experiments. If the NH_3 concentration is too low, the solution can be transferred to the stock bottle for dilute ammonia.

34

Hydrogen Peroxide and Blood

If one mixes blood with oxygen gas, the dark color of the former disappears, and it becomes a beautiful bright red. This is the transformation that the blood of a living animal undergoes in the course of breathing by drawing in oxygen, and through which at least a part of the animal warmth is maintained. If an animal is confined in oxygen, it continues breathing four times as long as in the same volume of atmospheric air. It is for this reason that this gas was once referred to as life-air. If the animal is subsequently removed, one finds that the blood in its veins is much redder than before, and if it has breathed oxygen gas for a long time the lungs are found to be in a kind of inflamed state. Those suffering lung disease therefore become significantly worse as a result of inhaling this type of air. Any matter that burns in air combines with the oxygen and increases in weight by as much as the weight of the consumed oxygen. Various bodies ignite (i.e., begin to combine with oxygen) at a temperature at which there is as yet neither glowing nor luminescence, and such bodies proceed to become oxidized without fire ensuing; indeed, such that they simply remain hot, although with the introduction of pure air fire does break out. This leads to a dual, differentiated state of combustion, one form of which occurs at the lowest possible heat, the other at the highest possible, whereby the products of the two can often be totally different. The same occurs if one allows any flammable type of gas to stream out into the air and then inserts a platinum wire into the stream, with the single distinction that gas types that are subject to very ready ignition ignite in this way within a few instants. Platinum is best suited to this experiment because it is sufficiently conductive to heat and is not itself oxidized. The experiment succeeds with other metals too, though less well and certainly; iron is closest to platinum in its behavior, but silver and gold appear to be too strongly conductive of heat for this purpose, thus diverting the heat. However, in all of these experiments the ignition and combustion is not a function exclusively of the temperature of the platinum. The metal itself, irrespective of the fact that it forms no compounds, participates in the process in a unique way. In the case of hydrogen I will come back to this material and introduce in detail that which our experience has taught us.

A body that has combined with oxygen is described as oxidized or combusted. The weight of such a body is equal to the sum of the weight of the *flammable body and the consumed oxygen. I noted above that we classify simple substances as electropositive and electronegative. Most of the latter form distinct compounds with oxygen that distinguish themselves through a sour [acidic] taste, and so have acquired the name acids. Electropositive substances, on the other hand, lead to compounds that are not acidic, having instead the property that they combine with acids in such a way as to destroy the acidic characteristics of the latter, so that these two classes of oxidized materials are as opposite to each other as the two electricities after which they are named. A very few electropositive substances form electronegative oxides (i.e., acids) with a certain amount of oxygen, as in the case of manganese, for example. As early as 1774 a French chemist,* Bayen, *made the observation that* Stahl's *teachings could not be applied to mercury, whose calcinate is subject to reduction without the addition of phlogiston, and that the calcination of mercury was due not to a loss of phlogiston but instead to combination with air, the added weight of which, together with the weight of the mercury, led to the weight increase the metal displayed during calcination.* Bayen's *experiments attracted* Lavoisier's *attention to this matter, and in the very same year he initiated his first experiments toward providing the absorption of air in the course of calcination. He melted some tin in a large glass flask sealed tightly against the atmosphere, the weight of which (as well as that of the tin) he had precisely ascertained in advance. After the tin had, over the course of three hours, acquired a thick coating of tin ash he allowed the apparatus to cool and weighed it. It had precisely the same weight as before. At approximately the same time,* Priestley *in England and* Scheele *in Sweden discovered the gas oxygen.* Scheele *demonstrated in a series of outstanding experiments the constitution of atmospheric air, and established the distinctiveness of nitrogen gas, oxygen gas, and carbon dioxide gas. He named the first of these* Aër mephiticus, *the second* Aër vitalis, *and the last* Acidum aëreum.*

In the year 1777 he published in Leipzig, in German language, his outstanding treatment of fire and air, in which all of his experiments were described. He took the risk in the process of proposing a theory of combustion. He had shown through experiments that phosphorus, which undergoes combustion in oxygen, absorbs this gas in its entirety, so that the flask in which the combustion took place became devoid of air, whereas under water the phosphorus would have combined with oxygen and released light and heat: light in the case of more phlogiston, heat in the case of less. But in all these experiments Scheele *had neglected to weigh the combusted material. Had this not been the case,* Scheele *would undoubtedly have been the originator of the oxidation theory. In the same year* Lavoisier *proved that combustion consisted of an absorption of oxygen, and that the material subsequent to combustion weighed as much more as the weight of the con-*

sumed oxygen. In the following year he showed that oxygen is a component of all acids, for which reason he named it Oxygène *(i.e., acid-former). Thereafter he revised the designations of fire air and air of life to* Gas Oxygène. *Lavoisier's much more comprehensive spirit conferred upon chemistry an entirely new turn. In 1789 he published his* Traité élémentaire de Chimie, *in which he presented the new teachings in all of their splendid contexts. One of the circumstances that caused the theory of oxidation to attract opposition was the uncertainty regarding the origin of the light accompanying combustion. Some suggested a middle course, in which they accepted the theory of oxidation, while regarding the light as a component of the flammable material, and giving it the name flammable, something that in the course of combustion binds with the heat of the oxygen gas and escapes in the form of fire. Some, who had no wish to accept oxidation, declared that phlogiston possessed a negative weight; i.e., that it strives to distance itself from the midpoint of the earth, whereby it causes material to which it is bound to become lighter, restoring to it its former weight once it has departed. I will deal later with how one explains these phenomena from an electrochemical perspective.*

Jöns Jakob Berzelius (1779–1848), Lehrbuch der Chemie *(1833) (from the translation by Friedrich Wöhler)*

A day will come in which zealous research over long periods of time will bring to light things that now still lie hidden. The life of a single man, even if he devotes it entirely to the heavens, is insufficient to fathom so broad a field. Knowledge will thus unfold only over the course of generations. But there will come a time when our descendants will marvel that we did not know the things that seem so simple to them. Many discoveries are reserved for future centuries, however, when we are long forgotten. Our universe would be deplorably insignificant had it not offered every generation new problems. Nature does not surrender her secrets once and for all.

Seneca, "Naturales quaestiones" Book 7 (1st Century A.D.)

Three types of Bombardier beetle live in Germany. These can grow to a length of two and a half centimetres. If the beetle feels threatened it expels a mixture of hydrogen peroxide and hydroquinone. Just as in the blood, an enzyme acts as a catalyst, the hydroquinone being converted to quinone and oxygen cleaved exothermically from the hydrogen peroxide. The oxygen serves as a propellant and shoots a jet of hot quinone at the attacker.

The beetle cannot spray the chemicals continuously, otherwise its body orifice would heat up; there are thus delays of fractions of a second between each ejection process.

Hydroquinone-Quinone

Apparatus

250-mL beaker, glass rod, safety glasses, protective gloves.

Chemicals

30 % hydrogen peroxide solution, blood.

Experimental Procedure

10 mL of blood are placed in the beaker and treated with 2 mL of the 30 % H_2O_2 solution. A violent reaction, accompanied by foaming, starts immediately; the blood is partially decolorised (bleaching action of H_2O_2), so that after the reaction the mixture resembles an ice cream sundae containing red fruit (colored figure 12).

Explanation

Enzymes can be divided into two main groups, the hydrolases and desmolases. The desmolases regulate the various oxidation-reduction processes. The desmolase which is present in blood and interacts with hydrogen peroxide is called catalase; it decomposes H_2O_2 into water and oxygen:

$$2\,H_2O_2 \xrightarrow{\text{catalase}} 2\,H_2O + O_2$$

The oxygen set free is a strong oxidant and decolorises the red hemoglobin.

35

Decomposition of Hydrogen Peroxide in the Presence of Manganese Dioxide

Name the greatest of all inventors: It is chance.

Mark Twain

What he saw was puzzling enough. Greek and Latin letters grouped with figures at various heights, interspersed with crosses and lines, placed above and below fractional lines, topped by other lines, equated by double lines, drawn together into massed formulae by large brackets. Single letters totally unintelligible to the layman circumvented both letters and numbers, while square root symbols preceded them, and numbers of letters hovered above and below them. Peculiar syllables, abbreviations of mysterious words were scattered everywhere, and among the necromantic columns were sentences and notes in ordinary language whose meaning was nonetheless far above the head of the average person, so that one could read them with no more understanding than if listening to an incantation.

Thomas Mann, "Royal Highness"

Safety Precautions

Safety glasses must be worn, as the reaction can be very violent!

Apparatus

500-mL Erlenmeyer flask, wooden splint, gas burner, safety glasses, protective gloves.

Chemicals

MnO_2, 10 % H_2O_2 solution.

Experimental Procedure

0.6 g of manganese dioxide are placed in the Erlenmeyer flask. 1–2 mL of the 10 % H_2O_2 solution are added slowly dropwise. The oxygen formed in the violent reaction which occurs can be detected with the help of a glowing wooden splint.

Explanation

Manganese dioxide catalyses the following decomposition:

$$2\ H_2O_2 \rightarrow 2\ H_2O + O_2$$

Waste Disposal

The manganese dioxide can be reused after drying.

36

Decomposition of Hydrogen Peroxide by Potassium Permanganate

It is surely more than merely probable that very diverse types and stages of decomposition exist, just as there are also very different types of bonding and relationship. What repels me from all chemical theories is that there is still no boundary line fixed between normal chemical operations and the hyperchemical operations of free nature. How do the countless plant juices arise from rain water? Horseradish alongside the water parsnip? And countless others as well. Are these merely decompositions of water? Where also then do the solid parts originate?

And if, in the end, different forms of air are created from water, who will assure me that this does not occur through hyperchemical operations?

Georg Christoph Lichtenberg

Safety Precautions

The reaction is extremely violent. A narrow-necked flask is **not** suitable! Safety glasses must be worn. The 30 % H_2O_2 solution is highly corrosive, so that all contact with the skin and eyes must be avoided. If contact does occur the affected areas must be washed under a stream of tap water for at least 10 minutes.

Apparatus

4-Liter flat-bottomed flask with a wide neck, 150-mL beaker, spatula, protective gloves, safety glasses.

Chemicals

30 % H_2O_2 solution, $KMnO_4$.

Experimental Procedure

100 mL of 30 % hydrogen peroxide are placed in the flat-bottomed flask. When 0.2 g of solid $KMnO_4$ are added there is a violent reaction, in which a large amount of water vapor and oxygen is formed in the flask.

Explanation

In this experiment the permanganate is reduced by the hydrogen peroxide to manganese(IV) dioxide, which can act as a catalyst.

Waste Disposal

The reaction mixture is allowed to stand until the gas evolution is complete. The aqueous solution containing a small amount of suspended manganese dioxide can be poured down the drain.

Decolorization of Permanganate Solution by Hydrogen Peroxide

We ... naturally hope that the world is orderly. We like it that way .... This idea of a basically ordered world is even one which, today, may be very important to us emotionally, may seem an important aspect of our salvation. All of us, including those ignorant of science, find this idea sustaining. It controls confusion, it makes the world seem more intelligible. But suppose the world should happen in fact to be not very intelligible? Or suppose merely that we do not know it to be so? Might it not then be our duty to admit these distressing facts?

This is a real difficulty. We are all children of the Enlightenment, whatever other forebears we may acknowledge. It has been a cardinal principle of our upbringing that we must never believe things simply because we want them to be true. But how are we to apply that principle to cases where our wanting-them-to-be-true is essentially a matter of the satisfaction of reason?[1]

Mary Midgley, "A Modern Myth and Its Meaning"

Safety Precautions

The H_2O_2 solution is highly corrosive. Avoid skin contact. Wear protective gloves and safety glasses.

Apparatus

250-mL beaker, glass rod, safety glasses, protective gloves.

Chemicals

Dilute $KMnO_4$ solution, 1 mol/L sulfuric acid, 30 % H_2O_2 solution, dilute sodium hydroxide solution.

Experimental Procedure

The dilute potassium permanganate solution in the beaker is acidified with the sulfuric acid and about 5 mL of the H_2O_2 solution are added. The decolorisation of the solution is not instantaneous but takes several minutes.

Explanation

Under these conditions Mn(VII) is reduced to Mn(II), the Mn^{2+} ions acting autocatalytically. In this case H_2O_2 acts as a reducing agent!

$$2\,MnO_4^- + 5\,H_2O_2 + 6\,H^+ \rightarrow 2\,Mn^{2+} + 8\,H_2O + 5\,O_2$$

Waste Disposal

The solution is neutralised with dilute sodium hydroxide solution and poured down the drain.

Reference

1 Cited by *Science*, **1995**, *269*, 567.

38

The Bleaching of Hair

But although all our knowledge begins with experience, it does not follow that it arises from experience.

Immanuel Kant, "Critique of Pure Reason" (1781)

A bleaching agent is a material that lightens or whitens a substrate through chemical reaction. The bleaching reactions usually involve oxidative or reductive processes that degrade color systems. These processes may involve the destruction or modification of chromophoric groups in the substrate as well as the degradation of color bodies into smaller, more soluble units that are more easily removed in the bleaching process. The most common bleaching agents generally fall into two categories: chlorine and its related compounds (such as sodium hypochlorite) and the peroxygen bleaching agents such as hydrogen peroxide and sodium perborate. Reducing bleaches represent another category. Bleaching agents are used for textile, paper, and pulp bleaching as well as for home laundering.

There is evidence of chemical bleaching of cloth prior to 300 BC. Soda ash prepared from the burning of seaweed was used to clean the cloth followed by souring, i.e., treatment with soured milk to neutralize the alkalinity remaining on the cloth. The cloth was then exposed to the sun to complete the bleaching process. Sun bleaching, which became known as crofting, occurred over a matter of weeks during which time the cloth was kept moist to enhance the bleaching process. During the eighteenth century improvements were developed including the use of sulfuric acid in the souring process and the use of lime in the cleaning process, though crofting still required large tracts of primarily coastal land. With the onset of mechanized weaving, the production of cloth was outstripping the availability of land, which set the stage of the introduction of chemical bleaching.

Scheele, a Swedish chemist, discovered chlorine gas in 1784 and demonstrated its use in decolorizing vegetable dyes. Berthollet first produced solutions of hypochlorite by combining chlorine gas with alkalies and suggested using the gas for bleaching. A Scottish bleacher followed the suggestion and introduced chlorine into a bleach works in Glasgow. The efficiency of the process lead to its widespread use, though the low pH resulted in fabric damage and worker health problems. Two chemists, Valette and Tennant, developed chlorinated lime solutions that minimized these difficulties.

Tennant received a patent in 1799 for bleaching powder formed by the absorption of chlorine gas by dry hydrate of lime. Although this eliminated the need for on-site manufacture of chlorine, evidence suggests its use by bleachers caught on slowly. The bleaching powder was the chief source of textile bleaches over the next century and was the impetus for much of the early chemical and chemical engineering developments. Tropical bleach was developed by the addition of quicklime to bleaching powder to make a material suitable for use under tropical conditions. After World War I, technology for shipping liquid chlorine and caustic economically was developed allowing for the on-site manufacture of sodium hypochlorite solutions at the textile mills. As a result, use of bleaching powder diminished.

After World War I, other chlorine-based bleaches were developed. In 1921 the use of chlorine dioxide for bleaching fibers was reported followed by the development of the commercial process for large-scale production of sodium chlorite. In 1928 the first dry calcium hypochlorite containing 70 % available chlorine was produced in the United States. This material largely replaced bleaching powder as a commercial bleaching agent.

Although hydrogen peroxide was prepared as early as 1818 by Thenard, the peroxides received little use as textile bleaches. Hydrogen peroxide was first prepared by the action of dilute sulfuric acid on barium peroxide, but later sodium peroxide and dilute acids were used. The prices of peroxides were high initially, and they found use only as a specialty chemical. Electrolytic methods in the 1920s allowed for the synthesis of less costly, strong (ca. 30 %) solutions of hydrogen peroxide. By 1930, hydrogen peroxide was being used to bleach cotton goods, wool, and silk on a limited scale. Shortly thereafter, the J-Box was developed by the FMC Corp. allowing for continuous bleaching of textiles with hydrogen peroxide. By 1940, 65 % of all cotton bleaching was done with hydrogen peroxide.

Thomas Mc Donough, "Bleaching Agents"[1]

Safety Measures

The H_2O_2 solution is highly corrosive. Avoid skin contact and wear safety glasses and protective gloves.

Apparatus

100-mL beaker, 250-mL beaker, glass rod, protective gloves, safety glasses.

Chemicals

30 % H_2O_2 solution, 2 mol/L ammonia solution, acetone, dilute hydrochloric acid.

Experimental Procedure

A lock of dark hair is first degreased in a 250-mL beaker containing a few mL of acetone. After drying, the hair is transferred to the 100-mL beaker; 50 mL of H_2O_2 and 10 mL of NH_3 solution are added. Care must be taken that the hair is covered by the liquid; during about an hour it becomes much lighter in color. It is removed, washed with water and dried.

Waste Disposal

The acetone can be collected and redistilled. The bleach solution is neutralised with dilute hydrochloric acid and poured down the drain.

Reference

1 Kirk-Othmer, *Encyclopedia of Chemical Technology*, 4th ed., vol. 4, John Wiley & Sons, Inc., **1992,** 271.

39

Invisible Inks

We re-enact with reverent attention
The universal chord, the masters' harmony,
Evoking in unsullied communion
Minds and times of highest sanctity.
We draw upon the iconography
Whose mystery is able to contain
The boundlessness, the storm of all existence,
Give chaos form, and hold our lives in rein.
The pattern sings like crystal constellations,
And when we tell our beads, we seve the whole,
And cannot be dislodged or misdirected,
Held in the orbit of the Cosmic Soul.

Hermann Hesse, "The Glass Bead Game"

Apparatus

Atomizer or sprinkler, 2 large beakers, brush, Erlenmeyer flask with stopper, absorbent paper or oatmeal wallpaper (white), safety glasses, protective gloves.

Chemicals

Ammonium thiocyanate, iron(III)chloride, potassium hexacyanoferrate(II), gallic acid, distilled water.

Experimental Procedure

10 g of the commercially available iron(III) chloride are completely dissolved in 200 mL of water in the Erlenmeyer flask. This solution is used to create the desired text or drawing on a large piece of absorbent paper with the help of a brush and the paper left to dry overnight. This placard, the size of which will depend on the size of the auditorium, is fixed to a board. If the latent image is sprayed with a solution of 2 g of NH_4SCN in 200 mL of water a red color is formed. Use of a solution of 5 g $K_4[Fe(CN)_6] \cdot 3\ H_2O$ in 200 mL of water causes the text or picture to be dark blue (colored figure 13). Treatment with gallic acid gives a black coloration. The colors can be varied slightly by varying the

salt concentrations. The same effects can naturally be obtained when the pla-
cards are painted with thiocyanate, hexacyanoferrate or tannin solutions and
sprayed with a solution of $FeCl_3$.

Explanation

In aqueous solution $FeCl_3$ and NH_4SCN form blood-red complexes such as
$[Fe^{III}(SCN)(H_2O)_5]^{2+}$ or the simple $Fe(SCN)_3$. With $K_4[Fe^{II}(CN)_6]$ the deep blue
complex $Fe_4[Fe(CN)_6]_3$ is formed, the "Prussian Blue" which has been known
since 1704. Its deep color is due to electron transfer between Fe^{II} and Fe^{III}.
Discrete Fe^{2+} and Fe^{3+} species can be detected in the crystalline solid
$Fe_4[Fe(CN)_6]_3 \cdot 14{-}16\ H_2O$.[1] $FeCl_3$ solutions form black complexes with gallic
acid, $C_6H_2(OH)_3COOH$, which is present in tannin.

Waste Disposal

After considerable dilution the solutions can be poured down the drain.

Reference

1 N. N. Greenwood, A. Earnshaw, *Chemistry of the Elements,* Pergamon Press, Oxford, New
   York, Toronto, Sydney, Paris, Frankfurt **1984**, 1271.

40

A Magic Box

Johann Christian Wiegleb writes as follows from Langensalza in Thuringia, where he is working as a chemist and apothecary, to the bookseller and writer *Friedrich Nicolai* in Berlin:

Langensalza, November 30, 1785

Your right honorable sir, I sent you a few days ago the drawings for the new copperplate engravings. That which I can utilize from the second portion ... I will use. Your suggestion to include a few sheets with invisible inks is a very good one, provided only that the requisite cobalt dye is available. I have absolutely no time for that, but I suggest to you for the preparation of the inks Mr. Hermbstaedt, *who is currently staying with Mr.* Wegelin *on the island.* H. Klaproth *would also be able to be of assistance. The green invisible ink is described in the 1st vol. of the* Natural Magic, *the red and blue is described in the* Chem. Annal. *of 1785, issue 7 and 8, p. 130. One could of course write almost anything on paper with each, but it would be nicer if, e.g., on one page a barren tree were to be drawn in the copper so that its barren branches could be provided with leaves, the color of which would become visible through heat. The blue ink could be applied to the colorless internal reaches of a male figure whose garment is suggested simply by a black outline in order that a blue garment would present itself upon warming.*

In a similar way the red could also be introduced; although the latter does not always show up in the desired quality ...

As soon as you are certain that the desired sheets can be prepared, I ask that you send me a message, so that I can mention it in a suitable place and describe the usage.

I remain with highest regard
your right honorable
obedient servant
Wiegleb

Apparently it was too difficult to provide Wiegleb's book with copper plate engravings colored with "sympathetic" inks.[1]

Apparatus

Cardboard carton, fine brush, pen with steel nib, crystallization dish, 2 sheets of paper, safety glasses, protective gloves.

Chemicals

$CuSO_4 \cdot 5\,H_2O$, 5 % ammonia solution, distilled water.

Experimental Procedure

10 g of copper sulfate are finely ground and dissolved in 100 mL of water. This light blue ink is used to draw a picture (in this case a zebra), on one of the sheets of paper with the brush and on the other with the steel-nibbed pen. The figures are invisible once they have dried. A crystallizing dish containing the 5 % ammonia solution is placed inside the cardboard carton, which is then covered by the lid in which a slit has been cut. The sheet for paper which has been colored with the brush is now pushed into the slit and left for 5 minutes inside the carton; on removing it the zebra is visible. It is blue, while if the sheet of paper for which the steel nib was used is treated in the same way the zebra appears in the colors blue and red-brown.[2]

Drawing of a zebra before and after treatment with ammonia

Explanation

The ammonia gas which evolves from the dish, the smell of which is easily recognised, forms the deep blue complex $[Cu(NH_3)_4]^{2+}$ with the Cu^{2+} ions. Metallic copper is deposited on the paper and the nib itself when the steel pen is used, the Cu^{2+} ions oxidizing the iron to Fe_{aq}^{2+} ions.

Waste Disposal

The solutions are neutralized with dilute acid, then treated with milk of lime until slightly basic and allowed to stand for a time. The mixture is now decanted, the $Cu(OH)_2$ precipitate transferred to the container used for collecting less toxic inorganic waste and the supernatant solution poured down the drain.

References

1 O. Krätz, *Historisch-chemische Versuche*, Aulis-Verlag, Köln, **1987**, 52.
2 F. Cherrier, *Chemie macht Spaß*, II. Teil, S. 30, Verlag J. F. Schreiber, Esslingen, Österr. Bundesverlag, Wien, Schwager und Steinlein, Nürnberg.

41

A Weather Station

The use of "sympathetic" inks reached a peak in the 18th century. These are solutions of dyestuffs which were said to "exert secret effects"; the most famous is based on cobalt dichloride. They were for example used in the "barometer flower" which is described as follows by a contemporary writer:[1]

... The anhydrous salt is obtained as a blue mass from the water-containing red chloride [$CoCl_2$] by gentle heating (to above 140 °C) in a downward-directed test tube or a porcelain dish, and upon cooling in moist air it gradually becomes red with the absorption of water; when moistened with water the color change occurs at once.

The anhydrous chloride is prepared by heating, and it dissolves readily in alcohol with a deep blue color. A saturated blue solution to which water is carefully added (to be sprayed from a wash bottle) becomes first violet and then, with more water, red. (According to A. Winkler, alcoholic cobalt chloride solution can be used for an approximate colorimetric water determination).

A very concentrated red chloride solution turns blue-violet upon heating.

Red cobalt chloride solution becomes deep blue if one introduces a sufficient amount of concentrated hydrochloric acid; ...

Handwriting applied to paper with the aid of a goose-quill feather in the form of a moderately dilute chloride solution appears upon drying to be pale red or even invisible if pink paper is utilized, but it shows up clearly with a blue color upon gentle warming (e.g., over a Bunsen burner).

Filter paper strips or pieces of white cotton dipped in moderately concentrated, aqueous, red cobalt chloride solution and dried in the air display a rose-red color if the air is moist, but become blue upon standing in dry air; with a moderate moisture content in the air the strips appear violet, and can thus serve as a crude hygrometer. With warming (e.g., high above a gas flame) the red color of the strip turns immediately into blue.

The color change is best demonstrated by placing the red paper or cloth strips or so-called barometer flowers in a desiccator, where they very quickly become blue. If a strip that has turned blue is spotted with moist blotting paper it immediately becomes red; breathing on it causes the blue color to change to violet. If steam from (gently) boiling water in a test tube is allowed to pass over a prepared strip, the side of the strip facing the steam

becomes red due to the moisture, whereas the external side becomes deep
blue as a consequence of warming ...

Apparatus

10-mL bottle with ground glass neck, 200-mL beaker, glass rod, brush, funnel,
gauze, sheet of white cardboard, safety glasses, protective gloves.

Chemicals

Gum arabic, $CoCl_2 \cdot 6 H_2O$, cooking salt, water-insoluble dyes, distilled water.

Experimental Procedure

The gum arabic is dissolved in 30 mL of hot water in the 100-mL bottle; the latter is stoppered and left to stand for 48 hours. The solution is then filtered
through a piece of gauze. A second solution is prepared consisting of 30 g of
$CoCl_2 \cdot 6 H_2O$, 15 g of cooking salt and 80 mL of water. This solution is also
filtered and mixed with the first solution. The liquid thus obtained is spread on
to the cardboard using the brush and allowed to dry out. The processes are then
repeated until the layer is sufficiently thick. The cardboard is now stood in the
open air, but not in direct sunlight. If it rains or the air is moist it turns pink.

We now mix exactly the same color using the water-insoluble dyes, paint a
sun on the cardboard and write the words "fine weather" next to it. This color
disappears when it is dry, as it is the same color as the background.

When the weather is fine the cardboard turns light blue. We now write the
word "rain" under the sun and make an appropriate drawing. The weather station is finished![2]

If the weather is fine the words "fine weather" appear in pink on a light blue
background. When it rains the word "rain" is visible in blue against a pink
background.

Explanation

In the presence of moisture the pink octahedral complex $[Co(H_2O)_6]^{2+} \, 2 \, Cl^-$ is
present. In dry air or on warming it loses water and is transformed into the blue
tetrahedral complex $[CoCl_4]^{2-}$. The blue color is brightened somewhat by the
addition of gum arabic and salt.

References

1 Cited by: O. Krätz, *Historisch-chemische Versuche*, Aulis-Verlag, Köln, **1987**, 55.
2 F. Cherrier, *Chemie macht Spaß*, II. Teil, J. F. Schreiber, Esslingen; Österr. Bundesverlag, Wien;
  Schwager und Steinlein, Nürnberg.

42

Crown Ether Inclusion Compounds

Amber, a resin with unusual inclusions: in his *"Metamorphoses"* the Roman poet *Ovid* (43 B.C. – 17 A.D.) describes the adventure of Phaethon, the son of the sun god Helios. Helios agreed to his son's request that he be allowed to drive the sun chariot. Phaethon had no experience, the horses ran wild, the chariot touched the heavens and the earth and left a track of fire in its wake. Zeus put an end to the chaos by means of a thunderbolt which catapulted Phaethon out of the chariot and down to the underworld Hades. Phaethon's sisters, the Heliads, wept for their brother; they were turned into dolls and their tears to amber.

   Amber is by no means a uniform substance, but a mixture of resins. Most of it was formed in the tertiary period, an era which began about 65 million years ago and ended about 2 million years ago. The resin exudated by the plants attracted insects which were trapped by the sticky mass. Thus the inclusion of these insects and plants gives palaeontologists the rare possibility of studying extinct life forms in their natural state. About a thousand types of insects have been discovered in amber, which is mainly found on beaches in East Prussia.[1]

Safety Measures

Chromates and dichromates are highly toxic and can be cancerogenic. Protective gloves should be worn! Trichloromethane is also highly toxic.

Apparatus

Ten large test tubes with stoppers, spatula, protective gloves, safety glasses.

Chemicals

Crown ether 18-crown-6, ethanol, trichloromethane, potassium permanganate, manganate and dichromate, sodium chromate and ammonium dichromate.

Experimental Procedure

0.03 g of one of the finely-powdered salts are placed in each of five test tubes, followed by 10 mL of trichloromethane. In five other test tubes the corresponding mixtures are treated with a few drops of a solution of the crown ether 18-crown-6 in ethanol. After shaking thoroughly there is no visible change in the test tubes 1–5 which do not contain the crown ether, i.e. the trichloromethane remains colorless. The test tubes 6–10 containing the crown ether show a different picture: in the test tube containing $KMnO_4$ the trichloromethane is dark violet in color while the others show the following colors: $(NH_4)_2Cr_2O_7$ very pale orange-yellow, $K_2Cr_2O_7$ deep orange, Na_2CrO_4 light yellow and K_2MnO_4 green. In the case of ammonium dichromate the color deepens somewhat when the test tube is heated for a short time.

Explanation

The most important and almost unique feature of the macrocyclic polyethers, the "*crown ethers*", is their tendency to form complexes with alkali metal ions. These complexes are held together by the Coulomb forces between the cation and the negative end of the C-O dipoles. The stability of these complexes, which often have a 1:1 stoichiometry between the alkali metal ion and the cyclic ligand, depends mainly on the efficiency with which the ion can be fitted into the crown ether ring; it also depends on the charge density of the ion and, as we show in this experiment, on the solvation power of the solvent. It is possible to prepare a large variety of crown ethers by varying the crown ether skeleton, for example by substituting a part of the ring by a phenyl group or by changing the number of oxygen atoms (fig.). In our example the deep color of the anions of the complexes of the potassium salts in trichloromethane demonstrate their remarkable stability. K^+ ions in fact appear to occupy a unique position within the series of singly charged positive ions. Through the pioneering

18 - crown - 6   12 - crown - 4

dibenzo - 14 - crown - 4   benzo - 15 - crown - 5   dibenzo - 18 - crown - 6

Structures of some crown ethers

work of *C. J. Pedersen* (1967) the crown ether complexes became one of the highlights of chemistry.[2,3]

Waste Disposal

The trichloromethane solutions are transferred to the containers used for collecting halogenated organic solvents, while the small amounts of undissolved salts should be placed in the container used for storing toxic inorganic waste.

References

1 A. Bachofen-Echt, *Der Bernstein und seine Einschlüsse*, Springer-Verlag, Wien, **1949.**
2 C. J. Pedersen, *J. Am. Chem. Soc.*, **1967,** *89,* 7017.
3 C. J. Pedersen, H. K. Frensdorff, *Angew. Chem.*, **1972,** *84,* 16; *Angew. Chem. Int. Ed., Engl.* **1972,** *11,* 16.

43

Color Effects due to Ligand Exchange in Nickel Complexes

In the literary rhetoric of ancient times the imagery of *"color"* (as first introduced by Plato) develops in two different senses: on the one hand *chroma* and *color* denote the one-sided coloration of objective facts in speech, and in particular the glossing over of negative aspects, and on the other they refer to the *ornatus*, the ornamentation of language by means of verse and figures of speech.

The Latin word *color* in the sense of *ornatus* is for example often used by *Cicero* in the course of polished *ut-pictura-poesis* imagery. A quote from *"Brutus"* will make this clear:

> *First of all let us have a look at the books of different authors, above all those of Cato. You then will understand that in his drawings there are missing only the floweriness and the colors of a kind of painting which was not known in those days.*

Cicero expresses similar sentiments in "De oratore": the beauty of the speech lies first of all in its color and freshness; however *Cicero* counsels us not to use *ornatus* too often, since in all things there is only a thin dividing line between the greatest delights and satiety.

Natural color change effects, in the original sense of the word, can be readily demonstrated by means of chemical reactions between nickel salts and aqueous ammonia solutions.

Apparatus

Five large test tubes with stand, dropping pipette, glass rods, flat porcelain dish, drying cupboard, powder bottle, porcelain spatula, safety glasses, protective gloves.

Chemicals

Crystalline nickel dichloride ($NiCl_2 \cdot 6\ H_2O$), 25 % NH_3 solution, absolute methanol or ethanol.

Experimental Procedure

Green nickel dichloride hexahydrate is kept at 120 °C in a porcelain dish in a drying cupboard until it has turned yellow; the product is at once transferred to a dried powder bottle and the latter stoppered. An approximately 0.5 mol/L aqueous solution of nickel dichloride is now prepared by adding 20 mL of water to 24 g of the salt in an Erlenmeyer flask. 4 test tubes are now filled two-thirds full with this solution. The contents of test tube 1 show the familiar green color of the nickel salt, test tube 2 is made turquoise by adding a few drops of ammonia solution, and test tube 3 dark blue by adding a larger amount of ammonia. An excess of ammonia is poured into test tube 4, yielding a violet coloration; a small amount of a pale violet precipitate may be formed from time to time. In another test tube a portion of the dehydrated salt is treated with alcohol, which takes on the yellow color of the salt. The colors now range from yellow via green and turquoise to blue and violet (colored figure 14). The colors in the test tubes can be varied between green and violet by adding either nickel dichloride or ammonia solution. A lime-green color rapidly appears if the alcoholic nickel dichloride solution is treated with ammonia vapor or allowed to come into contact with the moisture in the air.

Explanation

The yellow anhydrous nickel dichloride crystallises in a layered lattice in which the Cl-ions are arranged in almost the closest cubic packing. Alcohols cause the Ni-Cl bonds to be cleaved to only a small extent, but these are easily cleaved by water to give the green $[Ni(H_2O)_6]^{2+}$ cation. If ammonia solution is added, a stepwise ligand exchange between H_2O and NH_3 molecules occurs; the final step leads to the formation of the hexaammine complex $[Ni(NH_3)_6]^{2+}$, which is violet. The kinetically labile intermediates are of the composition $[Ni(H_2O)_m(NH_3)_n]^{2+}$. When water molecules are replaced by ammonia, which lies at the "stronger" end of the spectrochemical series, the (three) spin-allowed absorption bands of the octahedral nickel complex are shifted. Thus the aqua complex absorbs in the visible at $\lambda = 725$ nm, i.e. in the red region, so that the complex is bright green in color, while the hexaammine complex absorbs at $\lambda = 571$ (in the yellow-green range), thus giving rise to a blue-violet coloration. The strenght of this ligand field ($\Delta_0 = 10,750$ cm^{-1}) is much greater than that of the aqua complex ($\Delta_0 = 8,500$ cm^{-1}). This discussion of color effects applies only to octahedral 3d^8 systems with normal spin behavior, i.e. with magnetic moments $\mu_{eff} = 2.9–3.4$ Bohr magnetons (μ_B).

Waste Disposal

The solutions are treated with milk of lime and the precipitates transferred to the container used for collecting less toxic inorganic salts; the remaining liquid is neutralized and poured down the drain.

44

A Simple Separation of Cobalt and Nickel Salts

The mediaeval miners in the Saxonian Erzgebirge felt that they were being fooled by goblins they called "cobolds". Although some of the ores, known as *Bergkobalt*, appeared metallic they could not be smelted to give the expected metal, cobalt. These same minerals were however used as early as the third century in amounts as low as 0.3 % to give glass a blue color. Deep-blue cobalt glass itself is formed by fusing quartz sand with cobalt oxide and potassium carbonate. The melt obtained has the composition $K_2Co(II)(SiO_3)_2$. The darkest types, which can contain up to 7 % cobalt, are called cobalt blue, smalt and royal blue. *Leonardo da Vinci* used smalt to paint his *"Madonna in the Grotto"*. Only in 1735 was the Swedish chemist *G. Brandt* able to isolate metallic cobalt, which he called *"regulus cobalt rex"*, and in 1780 *T. O. Bergmann* recognised that the new metal was in fact an element.

Apparatus

250-mL separating funnel, stand with ring and boss, 100-mL measuring cyclinder, two 25-mL measuring cylinders, two 250-mL beakers, safety glasses, protective gloves.

Chemicals

$CoCl_2 \cdot 6\,H_2O$, $NiCl_2 \cdot 6\,H_2O$, 2-butanone, KSCN, distilled water.

Experimental Procedure

25 mL of a 1 % solution of $CoCl_2$ and of a 4 % solution of $NiCl_2$ are placed in the separating funnel. The almost grey solution is treated with 100 mL of 2-butanone, followed by about 0.6–0.7 g of KSCN. After brief shaking two phases separate: the upper one is deep blue, the lower one lime green.

Explanation

The aqua complex of the nickel(II) ion retains its green color and its octahedral structure in aqueous solution even in the presence of the SCN^- ions from the KSCN, which compete with the Cl^- ions; however, these SCN^- ions displace the water molecules from the pink $[Co(H_2O)_6]^{2+}$ cation, forming the tetrahedral violet $[Co(SCN)_4]^{2-}$ complex, which is quite stable in aqueous solution and dissolves in the organic phase to give a blue color. This complex shows a strong absorption band in the form of a doublet at $\lambda = 700$ nm and thus appears to be deep blue in color.

Waste Disposal

The mixture is shaken for a few minutes with milk of lime (pH 8). The precipitate is separated off and transferred to the container used for collecting less toxic inorganic salts, while the supernatant liquid is poured down the drain.

The Reaction of Iron(III) Chloride with Hydroxybenzenes

Color becomes fixed in bodies more or less permanently; superficially, or thoroughly.

All bodies are susceptible of color; it can either be excited, rendered intense, and gradually fixed in them, or at least communicated to them.

Chemical colors:

We give this denomination to colors which we can produce, and more or less fix, in certain bodies; which we can render more intense, which we can again take away and communicate to other bodies, and to which, therefore, we ascribe a certain permanency: duration is their prevailing characteristic.

Johann Wolfgang von Goethe, "Theory of Colors" (1810)

Apparatus

Seven large test tubes with stand, dropping pipettes, white cardboard, glass rods, safety glasses, protective gloves.

Chemicals

Catechol, resorcinol, hydroquinone, pyrogallol, phloroglucinol, salicylic acid, tannin (gallic acid), distilled water, ethanol, $FeCl_3 \cdot 6\,H_2O$.

Experimental Procedure

The seven test tubes are filled half full with a solution of 1 g of one of the hydroxybenzenes in 20 mL of a 1:1 mixture of water and ethanol. A small amount of a 5 % aqueous $FeCl_3$ solution is added to each test tube using a dropping pipette. The test tubes are shaken for a short time and the white cardboard placed behind them. The colors formed in the reactions are as follows:

1,2-Dihydroxybenzene (catechol):	green
1,3-Dihydroxybenzene (resorcinol):	dark violet
1,4-Dihydroxybenzene (hydroquinone):	first blue, then brownish yellow, after a time a green precipitate
1,2,3-Trihydroxybenzene (pyrogallol):	blood-red

1,3,5-Trihydroxybenzene (phloroglucinol):   blue
3,4,5-Trihydroxybenzoic acid (gallic acid):   black
2-Hydroxybenzoic acid (salicylic acid):   violet

Explanation

All hydroxybenzenes and their derivatives coordinate with the Fe^{3+} ions; the Fe(III)-O bridged complexes are generally polymeric and have different colors.

Waste Disposal

The residues are made slightly alkaline with a soda solution and transferred to the container used for storing halogenated organic solvents.

Five Colors from One Solution

Secret inks are no longer in fashion. Beggars no longer make their secret marks at the entrances to villages or great houses, and at the other end of the social scale the elite men's clubs of today no longer feel it necessary to code their messages with the help of the freemasons' alphabet. Even today, of course, secret messages are passed on with the help of the normal alphabet or with number combinations; spies can allow themselves this luxury, because it is not easy to break such number codes even with the help of complex mathematical models. In the laboratories of the secret services machines fight with one another: *Colossus versus Enigma.*

   In the olden days things were different. Thus one night in the year 538 B.C. writing appeared on one of the painted inner walls of the royal palace of Babylon; neither *Belshazzar* nor his minions could interpret it. But the young Jewish prisoner *Daniel* was able to read and interpret the writing on the wall. We can therefore be sure that the writer had used Hebraic letters. Thus whether a written code can be considered as secret writing depends on the recipient.[1] Be that as it may, it is possible to use simple chemical reactions to create a mysterious color effect, as in the following five-color experiment.

Apparatus

Five 250-mL beakers, four 50-mL beakers, five dropping pipettes, safety glasses, protective gloves.

Chemicals

1 % methanolic phenolphthalein solution, Na_2CO_3, $FeCl_3 \cdot 6\,H_2O$, NH_4SCN, $K_4[Fe(CN)_6] \cdot 3\,H_2O$, distilled water.

Experimental Procedure

1 mL of the phenolphthalein solution is added to 100 mL of water in the first beaker. The colorless solution is now poured into the second beaker, which contains 5 drops of a 5 % soda solution: it turns to a reddish violet. This liquid is now poured into the third beaker, which contains 5 drops of a 50 % $FeCl_3$ solution, and the color changes to a yellowish ochre. The fourth beaker contains 20 drops of a 30 % NH_4SCN solution, which causes a further color

change to blood-red when the liquid from the third beaker is poured in. Finally the contents of the fourth beaker are poured into the fifth, which contains 5 drops of a 5 % solution of $K_4[Fe(CN)_6]$: the result is a dark blue solution (colored figure 15).

Explanation

The indicator phenolphthalein, which is colorless in neutral solution, turns *reddish-violet* in the presence of the basic soda solution. In the third beaker the H_3O^+ ions set free in the hydrolysis of the iron(III) salt bind the OH^- ions from the soda solution, leading to decolorisation of the phenolphthalein; at the same time the solution turns *yellow* because of the presence of the hydrolysed iron(III) species (eqns. 1a, b):

(1a) $[Fe(H_2O)_6]^{3+} + H_2O \rightarrow [Fe(H_2O)_5(OH)]^{2+} + H_3O^+$

(1b) $[Fe(H_2O)_5(OH)]^{2+} + H_2O \rightarrow [Fe(H_2O)_4(OH)_2]^+ + H_3O^+$ etc.

In the fourth beaker iron(III) salts form deep *red* complexes such as $[Fe(SCN)(H_2O)_5]^{2+}$ with the SCN^- ions. The extreme stability of Prussian blue dominates in the fifth beaker, so that the *deep blue* color brings the series to a close. Deviations from the given concentrations can lead to slight differences in the effects due to the formation of precipitates or mixed colors.

Waste Disposal

The contents of beaker 5 should be diluted with water and poured down the drain.

Reference

1 H. Jochems, *Diagonal,* Journal of the Universität-GH Siegen, **1993**, *1*, 95.

47

Color Effects in Aqueous Systems Containing Divalent Metal Ions Derived from Selected 3d Elements

In the opinion of the Greeks, the world and all of life issued forth from water. Indeed, water was a god: Okeanos. Man later devised mermaids, elves, and nymphs. Man founded temples on the water shores. But man also built the cathedrals of Hildesheim, Paderborn, and Bremen on top of springs, and see: do not these cathedrals still exist? And are there not also water lovers and apostles of natural cures, whose souls have something almost peculiarly, gravely healthy about them? ... And naturally the man without characteristics also had modern knowledge somewhere in his consciousness ... And there is water: a colorless liquid, blue only in thick layers, odorless and tasteless, ... irrespective of the fact that bacteria, vegetable matter, air, iron, and sulfate and carbonate of lime are part of it as well in a physiological sense, and the prototype of all liquids is not a liquid at all, but a solid substance, a liquid, or a gas depending on the circumstances. Finally, the whole dissolves into a system of formulas that are somehow interrelated, and in the whole world there are only a few dozen men who think alike about such a simple thing as water; all others speak in languages that are at home somewhere between today and a few thousand years in the past. One must therefore say that if a person were simply to reflect a little he would in some sense find himself in an extraordinarily disorderly society!

Robert Musil, "Mann ohne Eigenschaften"

Apparatus

Eight large test tubes with stand, dropping pipettes, glass rods, protective gloves, safety glass.

Chemicals

Crystalline hydrates of the following sulfates: V(II), Cr(II), Mn(II), Fe(II), Co(II), Ni(II), Cu(II) and Zn(II), concentrated sulfuric acid, distilled water, *n*-heptane or *n*-hexane.

Experimental Procedure

1.5 g of each of the crystalline sulfate hydrates of manganese(II) to zinc(II) are dissolved in 10 mL of water and treated with 1 mL of concentrated sulfuric acid; these solutions are left to stand in the test tubes labelled 3 to 8. If the sulfates of V(II) and Cr(II) are available, then 1.8 g of each finely powdered salt is quickly added to 12 mL of 20 % sulfuric acid, to which a few zinc granules have been added, in the test tubes 1 and 2. Hydrogen evolution rapidly occurs, and when this is complete 5 mL of the hydrocarbons are poured carefully over the aqueous phase. If these salts are not available then 1.5 g NH_4VO_3 and 1.8 g $Cr_2(SO_4)_3 \cdot x\, H_2O$ respectively are added to 15 mL of 20 % sulfuric acid, a few zinc granules are added, and the solutions are treated with n-heptane or n-hexane as above. The test tubes 1 and 2 now contain the required solutions of VSO_4 and $CrSO_4$; the evolution of hydrogen can be caused to continue by adding further zinc granules or sulfuric acid. The following colors are observed in the test tubes 1 to 8: 1: lilac, 2: light blue, 3: pale pink, 4: light green, 5: pink, 6: emerald green, 7: lapis lazuli and 8: colorless.

The crystalline hydrates are generally speaking vitriols of the composition $M(II)SO_4 \cdot 7\, H_2O$. Like the stable manganese vitriol, copper vitriol crystallises with 5 molecules of water of crystallisation per metal ion. Mixed colors can appear if ammonium metavanadate is used as the starting material in test tube 1 because of the presence of several oxidation states of vanadium.

Explanation

The aqueous solutions contain hexaaqua complexes in which the metal ion lies at the center of an octahedron formed by the water molecules which act as the ligands; the ion-dipole interaction between the central ions and the oxygen donor atoms of the water molecules play the dominant role in stabilising the O_h symmetry of these complexes. The color of the aqua complexes is the result from transitions between ground and excited states of the 3d electrons of the central ion, which in the ligand field with its cubic symmetry also gives rise to typical absorptions in the visible region. Thus the V(II) complex absorbs at $\lambda = 570$ nm and itself appears lilac in color; the Cr(II) and Cu(II) complexes absorb at $\lambda = 715$ and 795 nm respectively and are thus blue. In these two cases an exact comparison with purely octahedral fields is no longer possible because of the Jahn-Teller effect ($3d^4$ and $3d^9$ systems!) The pink coloration of the cobalt(II) complex reflects the doublet absorption at $\lambda = 465$ and $\lambda = 515$ nm, while the nickel(II) solutions absorb at $\lambda = 400$ nm and $\lambda = 740$ nm and are themselves bright green in color. Zn(II) complexes, which are $3d^{10}$ systems, are colorless, while the pale pink coloration of the Mn(II) complexes is caused by several spin-forbidden transitions with low extinction coefficients: the $3d^5$ state is stable because it represents a half-filled shell. The light green color of the

Fe(II) salts cannot readily be explained: probably spin-forbidden transitions at $\lambda = 730$ nm are also responsible in this case. The field strength parameters can be calculated using the ligand field model; if the influence of the sulfate ions on the ligand field is neglected these are a measure of the stability of the aqua complexes in this series. The values of $\Delta_o \equiv 10$ Dq are as follows (in the sequence of the test tubes 1–7): 12,350 cm^{-1}, 13,900 cm^{-1*}, 7,800 cm^{-1}, 10,400 cm^{-1}, 9,300 cm^{-1} and 12,600 cm^{-1*}.[1]

Waste Disposal

The contents of the test tubes are stirred with milk of lime. The organic phase is separated and transferred to the container used for storing halogen-free organic solvents and the precipitate from the aqueous phase is placed in the container for less toxic inorganic waste, while the clear aqueous solution is poured down the drain.

Reference

1 C. K. Jørgensen, *Absorption Spectra and Chemical Bonding in Complexes,* Pergamon Press, Oxford, London, New York, Paris, **1962**, 284 et seq.

---

* tetragonally distorted and therefore not comparable.

48

Color Reactions as a Test for Solvents

Agite, Auditores ornatissimi, transeamus alacres ad aliud negotii! Quum enim sic satis excusserimus ea quatuor Instrumenta artis, et naturae, quae modo relinquimus, videamus quintum genus horum, quod ipsi Chemiae fere proprium censetur, cui certe Chemistae principem locum prae omnibus assignant, in quo se jactant, serioque triumphant, cui artis suae, prae aliis omnibus effectus mirificos adscribunt. Atque illud quidem Menstruum vocaverunt.

Hermannus Boerhaave (1668–1738)
De menstruis dictis in chemia, in: „Elementa Chemiae" (1733)[1]

Well then, my dear listeners, let us proceed with fervor to another problem! Having sufficiently analysed in this manner the four resources of science and nature, which we are about to leave (i.e. fire, water, air, and earth) we must consider a fifth element which can almost be considered the most essential part of chemistry itself, which chemists boastfully, no doubt with reason, prefer above all others, and because of which they triumphantly celebrate, and to which they attribute above all others the marvellous effects of their science. And this they call the solvent (menstruum).

Safety Precautions

Halogenated solvents and KCN are highly toxic. These experiments should only be carried out in a well-ventilated hood and not by schoolchildren and students!

Apparatus

Ten large test tubes with stand, hard white cardboard, glass rods, two 1-L beakers, gas burner, exsiccator, safety glasses, protective gloves.

Chemicals

Sulfuric acid, formic acid, acetic acid, ethanol, methanol, trichloromethane, dichloromethane, dimethyl formamide (DMF), dimethylsulfoxide (DMSO), dicyanobis(1,10-phenanthroline)-Fe(II), 30 % H_2O_2, solid NaOH.

Experimental Procedure

The test tubes 1 to 6 each contain 10 mL of the pure protic solvents H_2O, H_2SO_4, HCOOH, CH_3COOH, CH_3OH and C_2H_5OH, while numbers 7 to 10 contain the dipolar aprotic solvents $CHCl_3$, CH_2Cl_2, DMF and DMSO. A few crystals of the red-violet complex $Fe(phen)_2(CN)_2$ are powdered and added in small amounts to each of the solvents. The following colors are observed in the test tubes 1 to 6: brownish mauve, yellow, orange, bright red, wine-red and red-violet; the test tubes 7 to 10 show the colors blue-violet for trichloromethane, violet for dichloromethane and DMF and a dirty-brownish violet for DMSO (colored figure 16). The color effects can be seen more clearly against a white background. The color indicator does not dissolve completely in water (test tube 1) even on warming. It can if necessary be prepared as follows: 6.0 g of 1,10-phenanthroline are added to 3.9 g $Fe(II)(NH_4)_2(SO_4)_2$, Mohr's salt, in 400 mL of distilled water; the dark red solution is heated to just below its boiling point and a solution of 2 g KCN in 10 mL of water added (**Caution! KCN is very poisonous even in the smallest amounts!**) The mixture is stirred for a short time and allowed to cool and to stand for several hours. The precipitate is separated, washed with a little water and dried in the exsiccator.[2]

Explanation

The electrophilicity of a solvent, and thus its Lewis acid character, can be quantified in terms of the acceptor number AN.[3]

In this experiment the colors range from the yellow of concentrated sulfuric acid with AN = 130 via the yellow-orange of formic acid (AN = 84) to the red colors of acetic acid (AN = 53) and ethanol (AN = 37) to the blue-violet and violet colorations of the indicator with the weak Lewis acids trichloromethane, dichloromethane and dimethylformamide; though not completely homogeneous, the brown-violet color of the aqueous system (AN = 55) fits in to the scheme reasonably well.

Waste Disposal

The solutions should be stirred well with 30 % H_2O_2, first at pH 10–11, then at pH 8–9 (well-ventilated hood!) After about 15 minutes the halogenated solvents should be separated and transferred to the container used for organohalogens; the aqueous phase is neutralised and poured into the container used for collecting highly toxic inorganic waste.

References

1 Cited by *C. Reichardt, Solvents and Solvent Effects in Organic Chemistry,* 2nd Edn., Verlag Chemie, Weinheim **1988,** 1.
2 R. W. Soukup, *Chemie in unserer Zeit,* **1983,** *17,* 129.
3 U. Mayer, V. Gutmann, W. Gerger, *Monatsh. Chem.,* **1975,** *106,* 1235.

49

Equilibrium Reactions of Copper and Cobalt Complexes

As early as 1779 *C. L. Berthollet,* in his lecture *"Recherches sur les lois d'affinité"*, made reference to the influence of the masses of the substances reacting together and spoke of the "sphere of influence" within which the reaction must take place. However, the theoretical basis for a complete understanding of the influence of the active mass of a system of freely moving particles was not established until the second half of the 19th century, when the Norwegians *C. M. Guldberg* and *P. Waage* formulated the Law of Mass Action (1864–1867). In 1907 *G. N. Lewis* introduced the term "activity" to denote the number of particles in the sphere of influence (referred to by *Guldberg* and *Waage* as "volume concentration"); the activity is related to the concentration by the relation $a_i = f_a \cdot c_i$ ($f_a < 1$). This makes it possible to use the Law of Mass Action to deal with systems at higher concentrations.

In the case of reactions involving complex equilibria the influence of the interaction between ions and solvent molecules is generally characterised by means of large color differences.

Apparatus

Nine test tubes, stand, dropping pipettes, safety glasses, protective gloves.

Chemicals

1 mol/L $CoCl_2$ solution, 1 mol/L $CuCl_2$ solution, acetone, concentrated hydrochloric acid, 1 mol/L $AgNO_3$ solution, solid $NaNO_2$ and KSCN, concentrated ammonia solution.

Experimental Procedure

(I) 50 mL of the 1 mol/L $CoCl_2$ solution (24 g $CoCl_2 \cdot 6\ H_2O$ are dissolved in water and diluted to 100 mL) are divided equally between 5 test tubes. The pink solution in test tube 1 serves as a reference. A few drops of concentrated hydrochloric acid are added to test tube 2 until the color is deep blue, a few crystals of solid $NaNO_2$ are added to test tube 3 to give an orange color, and the contents of test tube 4 are rendered bright violet by adding solid potassium thiocyanate. When acetone is added dropwise to the solution in test tube 5 a blue layer is formed above the pink solution; addition of further acetone causes the whole contents of the test tube to take on this color. Finally, if the contents of test tube 2 are treated with an excess of a 1 mol/L $AgNO_3$ solution, a white precipitate is formed, the color of the liquid changing from blue to the original pink (colored figure 17).

(II) 50 mL of a 1 mol/L $CuCl_2$ solution (17 g $CuCl_2 \cdot 2\ H_2O$ are dissolved in water and diluted to 100 mL) are divided equally between four further test tubes. The color of the solution in test tube 6 is blue-green and on dilution with water approaches the sky-blue color of an aqueous copper sulfate solution. Addition of concentrated hydrochloric acid to test tube 7 leads to a color change to light green, which on warming becomes brown. Concentrated ammonia solution is added to test tube 8 until the color is deep blue; on dilution it becomes lighter. Test tube 9 is first treated with concentrated hydrochloric acid (like test tube 7); when an excess of a 1 mol/L $AgNO_3$ solution is added a white precipitate is obtained (see above) and the color changes to blue-green.

Explanation

(I) The equilibrium for the solutions of the cobalt(II) complexes is described by the following equation (eqn. 1):

$$(1) \quad [Co(H_2O)_6]^{2+} + 4\ Cl^- \rightleftarrows [CoCl_4]^{2-} + 6\ H_2O$$
$$ \text{pink} \text{blue}$$

The addition of hydrochloric acid and thus of Cl^- ions shifts the equilibrium to the right, which is shown by the blue color in test tube 2. In a somewhat modified manner acetone displaces water molecules from the coordination sphere and leads to the color effects observed in test tube 5. $NaNO_2$ forms an orange-yellow colored solution with Co_{aq}^{2+} ions (test tube 3), while the addition of KSCN leads to the formation of the violet complex $[Co(SCN)_4]^{2-}$. The Ag^+ ions react with the Cl^- ions in test tube 2 to give solid silver chloride, which is precipitated from the system, so that the original color of the $[Co(H_2O)_6]^{2+}$ complex is reformed.[1]

(II) The following equation probably describes the equilibrium in the case of the copper chloride solution (eqn. 2):

$$(2) \quad [Cu(H_2O)_4]^{2+} + 4\ Cl^- \rightleftarrows [CuCl_4]^{2-} + 4\ H_2O$$

Dilution of the solution in test tube 6 leads to the formation of the fully hydrated Cu^{2+} ion. Just as in the case of $CuSO_4$ or $Cu(NO_3)_2$ this is surrounded by four H_2O molecules in the plane and two at the apices of a tetragonally distorted octahedron (D_{4h} symmetry). The addition of concentrated HCl to test tube 7 shifts the equilibrium to the right, so that yellow $[CuCl_4]^{2-}$ ions dominate. The amount of water is further reduced on warming, and the solution takes on the brown color of highly concentrated $CuCl_2$ solutions. The $[Cu(NH_3)_4]^{2+}$ complex is formed when NH_3 is added. NH_4^+ and OH^- ions are formed when this system is diluted in test tube 8; since these do not take part in the complexation, the color becomes lighter. The processes occurring in test tube 9 have already been described above.

Waste Disposal

The solutions are combined, stirred with milk of lime and left to stand for a time. After decantation the precipitate is transferred to the container used for collecting less toxic inorganic salts; the supernatant liquid is neutralised with dilute sulfuric acid and poured down the drain.

Reference

1 C. L. Ophardt, *J. Chem. Educ.*, **1980**, *57*, 453.

50

The Colors of the Rainbow

In the examination of colored appearances we had occasion everywhere to take notice of a principle of contrast: so again, in approaching the precincts of chemistry, we find a chemical contrast of a remarkable nature. We speak here, with reference to our present purpose, only of that which is comprehended under the general names of acid and alkali.

We characterised the chromatic contrast, in conformity with all other physical contrasts as a more *and* less; *ascribing the* plus *to the yellow side, the* minus *to the blue; and we now find that these two divisions correspond with the chemical contrasts. The yellow and yellow-red affect the acids, the blue and blue-red the alkalis; thus the phenomena of chemical colors, although still necessarily mixed up with other considerations, admit of being traced with sufficient simplicity.*

The principal phenomena in chemical colors are produced by the oxidation of metals, and it will be seen how important this consideration is at the outset. Other facts which come into the account, and which are worthy of attention, will be examined under separate heads; in doing this we, however, expressly state that we only propose to offer some preparatory suggestions to the chemist in a very general way, without entering into the nicer chemical problems and questions, or presuming to decide on them. Our object is only to give a sketch of the mode in which, according to our conviction, the chemical theory of colors may be connected with general physics.

Johann Wolfang von Goethe, "Theory of Colors" (1810)

Acid-base indicators are weak organic acids (denoted by the symbol IH) whose deprotonation is accompanied by a color change:

$$\text{IH} \quad \leftrightarrows \quad \text{I}^- + \text{H}^+ \qquad K_{\text{IH}} = \frac{c_{\text{H}^+} \cdot c_{\text{I}^-}}{c_{\text{IH}}}$$

(color 1)　　　(color 2)

The effect of base on phenolphthalein

name	color 1	color 2	pH interval for color change
m-cresol purple	red	yellow	1.2–2.8
thymol blue*	red	yellow	1.2–2.8
dimethyl yellow	red	yellow	2.9–4.1
bromophenol blue	yellow	blue-violet	3.0–4.6
congo red	blue	red	3.0–5.2
methyl orange	red	orange	3.1–4.4
bromocresol green	yellow	blue	3.8–5.4
methyl red	red	yellow	4.4–6.2
alizarin red	yellow	red-violet	5.0–6.6
bromothymol blue	yellow	blue	6.0–7.6
phenol red	yellow	red	6.4–8.2
thymol blue*	yellow	blue	8.0–9.6
phenolphthalein	colorless	red-violet	8.2–10.0

* thymol blue goes through two color changes

In practice mixtures of indicators are often used, for example the universal indicator which provides preliminary information on acidic (red), neutral (yellow to yellow-green) or basic (green and blue) behavior. In our experiment we shall also use some mixed indicators.

Apparatus

Six large test tubes with stand, hard white cardboard, long-stem funnels (glass or polyethylene, filter paper, two glass bottles, safety glasses, protective gloves.

Chemicals

Indicators or indicator mixtures (see below), 0.002 mol/L hydrochloric acid, 0.01 mol/L NaOH, ethanol.

Experimental Procedure

50 mg of the following indicators are introduced into the six test tubes via the long-stem funnels: phenol red (test tube 1), methyl red-HCl and phenolphthalein in the ratio 5:1 by weight (2), methyl red-HCl (3), methyl red-HCl and methylene blue in the ratio 4:1 by weight (4), brilliant green and *m*-cresol purple in the ratio 5:1 by weight (5) and bromophenol blue and alizarin red in equal amounts (6).

7 mL of a solution prepared from 40 mL of 0.002 mol/L hydrochloric acid and 10 mL of ethanol are added to each tube. The following colors are obtained (from 1 to 6): orange, pink, red, blue, blue-green, yellow. If these solutions are treated with an equal amount of 0.01 mol/L NaOH and stirred, the *colors of the rainbow red, orange, yellow, green, blue and violet* appear in this order. The white cardboard makes the colors appear more brilliant if held behind the test tubes (colored figure 18).

Explanation

The addition of the sodium hydroxide, which is five times as concentrated as the hydrochloric acid, causes the solution to change from being acidic to being basic; the indicator mixtures are chosen so that the colors of the rainbow are formed.[1] (The indicators methylene blue and brilliant green help to vary the color palette).

A detailed description of a number of experiments concerned with the way in which acid-base indicators work can be found in the literature.[2]

Waste Disposal

The solution can be poured down the drain.

References

1 R. E. Loffredo, D. Crookston, *J. Chem. Educ.*, **1993**, *70*, 774.
2 B. Z. Shakhashiri, *Chemical Demonstrations, a Handbook for Teachers of Chemistry*, University of Wisconsin Press, Madison, London, **1989**, *3*, 41.

51

Plant Dyes
as Universal Indicators

The colors of organic bodies in general may be considered as a higher kind of chemical operation, for which reason the ancients employed the word concoction, πέψις, to designate the process. All the elementary colors, as well as the combined and secondary hues, appear on the surface of organic productions, while on the other hand, the interior, if not colorless, appears, strictly speaking, negative when brought to the light. As we propose to communicate our views respecting organic nature, to a certain extent, in another place, we only insert here what has been before connected with the doctrine of colors, while it may serve as an introduction to the further consideration of the views alluded to: and first, of plants.

Flowers of the same genus, and even of the same kind, are found of all colors. Roses, and particularly mallows, for example, vary through a great portion of the colorific circle from white to yellow, then through red-yellow to bright red, and from thence to the darkest hue it can exhibit as it approaches blue.

A process somewhat similar takes place in the juicy capsule of the fruit, for it increases in color from the green, through the yellowish and yellow, up to the highest red, the color of the rind thus indicating the degree of ripeness. Some are colored all round, some only on the sunny side, in which last case the augmentation of the yellow into red, – the gradations crowding in and upon each other, – may be very well observed.

Many fruits, too, are colored internally; pure red juices, especially, are common.

Johann Wolfgang von Goethe, "Theory of Colors" (1810)

The blue dye *lacca*, which can be extracted from the lichens *Roccella* and *Lecanora* with alkali and milk of lime, is the basis for the well-known litmus solution, which functions as an acid-base indicator in the pH range 4.4–8.0. The name litmus comes from the Dutch *lackmoes* (from moes = mush, paste).

The litmus solution, which is violet in neutral solution, turns blue in the presence of bases and red when acids are added. The main component of litmus is a polymer containing 7-hydroxy-2-phenazinone chromophores. Other natural products can also be used as indicators for acid-base reactions, though they are not as specific. Thus beetroot juice is purple in the presence of acetic acid and remains so up to pH 7; on the addition of ammonia it turns blue-violet and at pH 12 it becomes brown, flocculation being observed. In this experiment we shall use an aqueous extract of red cabbage, which is a much better indicator.

Apparatus

Eight large test tubes with stand, glass rods, safety glasses, protective gloves.

Chemicals

Aqueous extract of red cabbage, 0.1 mol/L solutions of HCl (I), KH_2PO_4 (II), Na_2HPO_4 (III), K_3PO_4 (IV), NaOH.

Experimental Procedure

The test tubes 1 to 8 are filled half full with freshly prepared and filtered aqueous extract of red cabbage. In test tube 1 this is diluted with water, while the test tubes 2 to 7 are treated with 5 mL of the following buffer solutions and the mixture stirred with a glass rod:

test tube no.	composition of the buffer	pH value	color
1	–	ca. 7	violet
2	9.5 mL I + 0.5 mL II	2.1	red
3	0.5 mL I and 9.5 mL II	3.6	red-violet
4	9.0 mL II + 1.0 mL III	5.9	violet
5	4.0 mL II + 6.0 mL III	7.0	blue-violet
6	2.0 mL II + 8.0 mL III	7.4	blue
7	5.0 mL II + 5.0 mL IV	9.8	blue-green

10 mL of NaOH are added to test tube 8: the solution turns green at pH > 12 (colored figure 19).[1]

Explanation

Many berries, blossoms and flowers as well as red cabbage contain *anthocyanins*, the intense color of which hides the green hue of the chlorophyll in the leaves. Cyanidine chloride, which was first prepared in a pure form by R. Willstätter in 1913, can be considered to form the basis of the *anthocyanins*, in which the color component is linked to glucose or rhamnose via a glycosidic bond. The name *anthocyanin* comes from the Greek words *kyanaeos*, blue and *anthos*, color.

Waste Disposal

The solutions are neutralized and poured down the drain.

Reference

1 F. Bukatsch, O. P. Krätz, G. Probeck, R. J. Schwankner, *So interessant ist Chemie*, Aulis-Verlag, Köln, **1987**, 41.

52

Chemical Equilibria in Mineral Water

Mineral waters stimulate the appetite, improve the bowel functions, promote digestion, act as diuretics and, naturally, still thirst; these are real wonder drugs which are produced in the depths of the earth. They contain bicarbonates and valuable trace elements and are normally freed from iron and treated with carbon dioxide before being put on the market. A well-chilled mineral water works wonders on the "morning after"!

Apparatus and Chemicals

Full bottle of mineral water (no natural or artificial colorings!) with screw top, dropping pipette, 0.1 % alcoholic solution of bromocresol green, safety glasses.

Experimental Procedure

The closed bottle is cooled to around 0 °C (either before the demonstration in a refrigerator or during it in an ice bath). The bottle is opened and about a third of its contents quickly poured out. Three drops of the bromocresol green solution are added to the remaining contents and the bottle re-closed. A clear yellow color is observed. The bottle is now left to stand for about 10 minutes at room temperature, shaken vigorously and opened to release the CO_2 pressure built up. This process is repeated until practically no pressure buildup is noticeable. The color of the mineral water is now green. The open bottle is now heated for a few minutes in a water bath containing hot water, after cooling the solution should be deep blue.

Explanation

Bromocresol green is an indicator acid with a pK_A of 4.7; it colors solutions yellow below pH 3.8. This is due to the bicarbonate formed when carbon dioxide dissolves in water; on cooling and at a sufficient pressure this affords a sufficient quantity of H_3O^+ ions (eqns. 1 and 2):

$$(1) \quad CO_2(g) + H_2O \rightleftharpoons CO_{2\,aq}$$

$$(2) \quad CO_{2\,aq} + 2\,H_2O \rightleftharpoons H_3O^+ + HCO_{3\,aq}^-$$

Label of a bottle of mineral water

A considerable amount of the carbon dioxide can escape when the bottle is opened, so that the pH lies between 3.8 and 5.4 and a green color is formed. The remaining gas can be displaced by vigorous heating of the contents of the bottle; the pH value increases further and the color of the indicator changes to an intense blue.[1]

Reference

1 C. A. Snyder, D. C. Snyder, *J. Chem. Educ.*, **1992**, *69*, 573.

53

Dry Ice and Indicators

Effervescent powder contains a mixture of tartaric acid and Bullrich salt, substances which, although related, combine under normal conditions only slowly, but if one introduces water into the mix they rush into each others arms with a triumphant effervescence.

Johann Wolfgang von Goethe

Apparatus

Four 800-mL beakers, glass rods, two 5-mL measuring cylinders, safety glasses, protective gloves.

Chemicals

Phenolphthalein, methyl red, phenol red, Yamada's universal indicator (see solution 4 below), NaOH, dry ice, distilled water.

Solution 1

0.05 g of phenolphthalein are dissolved in 50 mL of ethanol and the solution diluted to 100 mL with H_2O.

Solution 2

0.02 g of methyl red are dissolved in 50 mL of ethanol and the solution diluted to 100 mL with H_2O.

Solution 3

0.04 g of phenol red are dissolved in 11 mL of 0.1 mol/L NaOH and the solution diluted to 100 mL with H_2O.

Solution 4

0.005 g of thymol blue, 0.012 g of methyl red, 0.06 g of bromothymol blue and 0.10 g of phenolphthalein are dissolved in 100 mL ethanol. 0.01 mol/L NaOH is added until the solution is green, and the solution is finally diluted to 200 mL with H_2O; this is *Yamada's universal indicator*.

Experimental Procedure

Each beaker contains 600 mL of distilled water. Five mL of one of the above indicator solutions are added to the beakers, followed by 5 mL of 0.1 mol/L sodium hydroxide. Pieces of dry ice, about the size of a walnut, are now added to the colored alkaline solutions. A reaction at once occurs, the liquids appear to boil, and the colors change as follows:

indicator	color change	pH range
phenolphthalein	red-violet → colorless	10.0–8.2
methyl red	yellow → red	6.2–4.4
phenol red	red → yellow	8.2–6.4
Yamada's universal indicator	violet → blue → green → yellow → orange → red	10.0–4.0

Explanation

The reaction between carbon dioxide and water leads to the formation of bicarbonate and protons, so that the pH is lowered and the color of the indicator changes (colored figures 20, 21).

$$CO_2 + 2\,H_2O \rightleftarrows HCO_3^- + H_3O^+$$

The excess of CO_2 shifts the equilibrium to the right, so that the pH value of the solution remains relatively constant. Solid carbon dioxide was obtained for the first time by *Thilorier* in 1834.[1]

Waste Disposal

The solutions can be poured down the drain.

Reference

1 M. Thilorier, *Ann. Chim. Phys.*, **1835**, *60*, 432.

54

Self-Organization in Solution

Like the artist, the chemist engraves into matter the products of creative imagination. The stone, the sounds, the words do not contain the works that the sculptor, the composer, the writer express from them. Similarly, the chemist creates original molecules, new materials and novel properties from the elements provided by nature, indeed entire new worlds, that did not exist before they were shaped at the hands of the chemist, like matter is shaped by the hands of the artist, as so powerfully rendered by Auguste Rodin.

Indeed chemistry possesses the creative power as stated by Marcelin Berthelot "La chimie crée son objet" ("Chemistry creates its object").[1]

Jean-Marie Lehn

The convection currents in a liquid can best be made visible by making use of patterns caused by a color change. This effect can be achieved very simply when a slightly alkaline aqueous surface which is colored by means of an indicator is exposed to an atmosphere of HCl. The convection is set in motion by the warmth provided by the lamp of an overhead projector. A variety of patterns can be obtained by changing the starting conditions slightly.

Apparatus

Overhead projector, glass dish (diameter 10 cm, height 1.5 cm), glass dish (diameter 15 cm), filter paper (diameter 14 cm), paper towels, 250-mL beaker, 1-mL pipette, safety glasses, protective gloves.

Chemicals

1 % bromocresol green indicator in ethanol solution, half-concentrated HCl, 0.005 mol/L sodium hydroxide solution (100 mL).

Experimental Procedure

About 100 mL of the 0.005 mol/L NaOH solution are placed in the beaker and 1 mL of the 1 % bromocresol green indicator solution added. 50 mL of the resulting solution are poured into the smaller glass dish which is standing on the overhead projector. The picture projected on to the wall (or screen) shows a blue solution. A filter paper is impregnated with half-concentrated hydro-chloric acid contained in the larger glass dish, dried slightly between paper towels and placed for about 10 seconds on the glass dish containing the blue solution. After the filter paper is removed a chemical pattern begins to appear, yellow threadlike bands on a blue background which slowly increase in size (colored figures 22).[2]

Explanation

The blue alkaline solution absorbs HCl at its surface and changes color to yellow. Convection currents cause the warmer blue solution to rise constantly from below to the surface and to displace the yellow surface layer.

Waste Disposal

The solution can be used for further experiments, but must first be regenerated by the addition of 1–2 drops of 0.1 mol/L NaOH. If it is not required, it can be poured down the drain because of its high dilution.

References

1 J.-M. Lehn, *Supramolecular Chemistry*, VCH, Weinheim, **1995**, 206.
2 P. G. Bowers, L. J. Soltzberg, *J. Chem. Educ.*, **1989**, *66*, 210.

55

Acidic and Basic Salts

Even in ancient times natural soda from the lower Egyptian soda lakes and plant soda (which was isolated by leaching the ash obtained by burning plants which grew on the seashore) were used for cosmetic purposes, for cleaning and bleaching clothing and last but not least for the manufacture of glass. Just like potash (potassium carbonate, which was obtained by evaporating wood ash lye in "pots"), soda was used as a mild base. On the other hand these substances are typical salts as obtained in acid-base reactions. The French scholar *G. F. Rouelle* (1703–1770), professor at the Botanical Gardens in Paris and elected a member of the Paris Academy of Sciences in 1744, distinguished in his publications on the nature of salts (1744–1754) between *"... middle or neutral salts, in which the base is exactly neutralized by the acid ..."* (such as cooking salt or Glauber's salt) and the acidic or basic salts, which obtain their specific properties from the strength of the acidic or basic component. The acidic salts include hydrates magnesium(II) and iron(III) chloride, while soda, potash and the alkali metal salts of the plant acids are basic.

Apparatus

Two 100-mL round-bottomed flasks with stoppers, two large test tubes, test tube stand, stand with fixing device, gas burner, ice bath, safety glasses, protective gloves.

Chemicals

$CH_3COONa \cdot 2\ H_2O$, $MgCl_2 \cdot 6\ H_2O$, alcoholic phenolphthalein solution, methyl red, distilled water.

Experimental Procedure

(I) 3 g of sodium acetate are placed in one of the test tubes, 10 mL of distilled water and 5 drops of an alcoholic phenolphthalein solution are added and the mixture shaken: a red-violet coloration is observed. 15 g of the acetate are now placed in the round-bottomed flask, 5 drops of the phenolphthalein solution are added, the flask stoppered and carefully heated. After a few seconds the salt melts in its water of crystallisation and the contents of the flask turn raspberryred. On cooling in the ice bath the color disappears again.

(II) 3 g of $MgCl_2 \cdot 6\,H_2O$ are dissolved in 10 mL of distilled water in a test tube and treated with 5 drops of an alcoholic solution of methyl red: a bright red color is observed. 12 g of the magnesium salt are now placed in the second round-bottomed flask and 5 drops of the methyl red solution are added, the mixture turning yellow. The flask is stoppered and warmed. The salt quickly melts, the mixture turning bright red; on cooling in the ice bath the color changes back to yellow.

Explanation

(I) Sodium acetate is completely dissociated in aqueous solution. An equilibrium is formed between the acetate ions and the water molecules; because of the weak protolytic properties of acetic acid this is shifted to the right (eqn. 1):

$$(1)\ CH_3COO^- + H_2O \rightleftarrows CH_3COOH + OH^-$$

Phenolphthalein is thus colored red. However, the indicator cannot enter the crystal lattice of the solid acetate. Only when the salt is fused, do the ions become mobile and undergo the hydrolysis described by eqn. (1). Since the equilibrium constant is temperature-dependent, the hydroxide ion concentration increases with increasing temperature, so that the red coloration of the indicator increases. On cooling the original lattice is re-formed and the residue is colorless.

(II) $MgCl_2 \cdot 6\,H_2O$ is a Brønsted acid and reacts as follows in aqueous solution (eqn. 2):

$$(2)\ [Mg(H_2O)_6]^{2+} + H_2O \rightleftarrows [Mg(H_2O)_5(OH)]^+ + H_3O^+$$

The heating of the solid salt again leads to melting and dissolution in the water of crystallisation; hydrated Mg^{2+} ions and H_3O^+ ions are formed just as in the aqueous solution (eqn. 2), so that the red coloration typical of methyl red at $pH < 6$ is observed. In the crystal lattice the indicator does not change from its normal yellow color.

Waste Disposal

The solutions can be poured down the drain.

56

The Amphoteric Behavior
of Aluminum

The matter it selfe, of which the Universe doth consist, is of it selfe very tractable and pliable. That rationall essence that doth governe it, hath in it selfe no cause to doe evill. It hath no evill in it selfe neither can it doe any thing that is evill: neither can any thing be hurt by it. And all things are done and determined according to its will and prescript.

Doth either the Sunne take upon him to doe that which belongs to the raine? or his son Æsculapius that, which unto the Earth doth properly belong? How is it with every one of the starres in particular? Though they all differ one from another, and have their severall charges and functions by themselves, doe they not all nevertheless concurre and cooperate to one end?

Marcus Aurelius Antoninus, "His Meditations", 6, 1 and 38

Apparatus

1-L beaker, magnetic stirrer, hard white card, safety glasses, protective gloves.

Chemicals

6 mol/L NaOH, $KAl(SO_4)_2 \cdot 12\ H_2O$, concentrated sulfuric acid, distilled water, universal indicator solution.

Experimental Procedure

30 g of finely powdered alum, $KAl(SO_4)_2 \cdot 12\ H_2O$, are dissolved in 750 mL of water in the beaker and stirred for a few minutes. On addition of 5 mL of concentrated sulfuric acid a clear solution is obtained which shows an acidic reaction. Small amounts of a 6 mol/L NaOH solution are slowly stirred in until the solution becomes milky and $Al(OH)_3$ begins to precipitate. On further addition of sodium hydroxide solution the milkiness suddenly disappears and the solution becomes clear again. If a few drops of universal indicator are added at the beginning of the experiment, the complete color spectrum from wine-red via yellow-orange to deep blue is observed. The colors are particularly striking, if the experiment is carried out against a white background.

Explanation

The sulfuric acid solution of the alum is neutralised by NaOH until solid $Al(OH)_3$ is formed according to eqn. 1:

$$(1)\ [Al(H_2O)_6]^{3+} + 3\ OH^- \rightleftarrows Al(OH)_3 + 6\ H_2O$$

An excess of OH^- ions leads to the formation of the tetrahydroxoaluminate complex, which is very soluble in water, and related species according to eqn. 2:

$$(2)\ Al(OH)_3 + OH^-_{\ aq} \rightarrow [Al(OH)_4]^-,\ [Al(OH)_4(H_2O)_2]^-\ \text{respectively.}$$

Other polynuclear complexes linked by OH^- bridges are also formed.

Waste Disposal

After neutralization and dilution with water the solution can be poured down the drain.

The Ammonia Fountain

Establishing a very unique chemistry for vapors, wherein nothing is discussed other than the mixing of vapors, at most in conjunction with permanent elastic fluids. I believe without question that something good would come from this. Admittedly, the vapors must be employed not merely at the temperature at which they first readily form, but also at the very highest heat, and in general one would need to think as well about sundry variations of vapors.

Georg Christoph Lichtenberg

Experiments Pro and Contra Vacuum

Question from al-Biruni:

If we take it as a given that there is no such thing as emptiness, neither within the world nor outside it; how can it be that water forces its way into a bottle and then rises up inside the bottle when one sucks it out and then holds it inverted in water?

Answer from *Avicenna:*

This is not a result of emptiness; the phenomenon is due instead to the fact that in the course of sucking on the bottle, whereby the air cannot escape due to the nonexistence of emptiness, this sucking action sets the air in the bottle in continuous, powerful motion. The continuous powerful motions produce a warmth and a heat, and the heat produces an expansion in the air, and when the air in the bottle expands it demands more space. Therefore a portion of it necessarily comes out, and that which the bottle can hold remains therein.

Then when the cold of the water reaches it, it becomes more dense, and draws itself together and occupies less space. Since emptiness cannot exist, water forces its way into the bottle as a result of contact with the cold body, and to the same extent that the air expanded. Do you not see that you would achieve the same effect if you did not suck, but rather the opposite, namely that you blew into the bottle? You must blow into it uninterruptedly and continuously until the motion from blowing has caused the air in the bottle to become warm, at which point you turn the bottle upside down on the water. That is established. In the same way you can also achieve this effect if you heat the bottle, and that should be sufficient answer.

Reply from *al-Biruni:*

Your arguments serve only the advocates of the vacuum, because if sucking causes air to expand, as you have explained, and it comes out of the bottle, which can no longer contain it, where should it go if there is not emptiness anywhere in the world? Unless one were to insist that all of a sudden some-where in the world a corresponding volume of air were to be reduced and draw itself together, and that the contraction an expansion balanced out. But with respect to your words "that is established" I have tried it, and I have also carried out the opposite process. And in fact the air came gurgling out of the bottle, while no water whatsoever entered it, and the number of bottles that burst for me in the process would suffice for the water of the Amudarja.

Taken from: *Al-as'ila*

Safety Precautions

Ammonia vapor is highly corrosive. The flask should be filled in a well-ventilated fume hood. Protective gloves and safety glasses should be worn.

Apparatus

Two 2-L round-bottomed flasks without ground glass joints, two cork rings, stand, clamps, bosses, glass tube (50 cm long, diameter 7–8 mm, with a pointed end), rubber stopper with one hole, rubber stopper with two holes, pipette pump or safety bulb, glass tube of diameter 7–8 mm bent at right angles, pieces of rubber tubing for use as connectors, safety glasses, protective gloves.

Chemicals

1.6 L distilled water, 1 mL mixed indicator No. 5 (pH range 4.4–5.8), a few drops of dilute hydrochloric acid, ammonia from a gas bottle.

Experimental Procedure

One of the 2-L round-bottomed flasks is stood on a cork ring; 1.6 L of distilled water and 1 mL of the indicator solution are added, followed by a few drops of acid until the contents turn red. The flask is now attached to the retort stand with a clamp and the 50 cm glass tube inserted through one hole of the rubber stopper with 2 holes. The bent glass tube is inserted into the other hole.

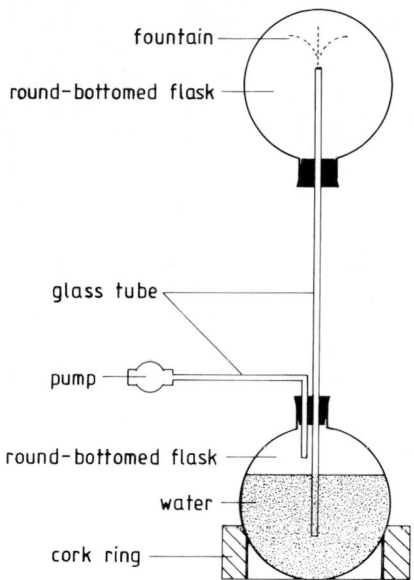

fountain

round-bottomed flask

glass tube

pump

round-bottomed flask

water

cork ring

Apparatus of the ammonia fountain

Ammonia is passed into the second, dry 2-L round-bottomed flask until the air has been completely displaced. This takes about a minute, or until the smell of ammonia is clearly noticeable. The opening of the flask should be pointed downwards. This flask is now carefully linked to the first flask as follows: the upper end of the long glass tube is passed through the rubber stopper with one hole, and this stopper is used to close the flask. The pointed end of the glass tube should be about 3 cm from the base of the flask, which is also attached to the stand by means of the ring clamp. The pump or safety bulb is now used to force air through the bent glass tube into the lower flask. The liquid rises in the glass tube towards the flask filled with ammonia. As soon as the water comes in contact with the ammonia the water is sucked into the upper flask with extreme force. The solution in this flask is colored green.

Explanation

One liter of water dissolves 1176 liters of ammonia at 0 °C and 702 liters at 20 °C.

Waste Disposal

The weakly basic solution is neutralized with a few drops of hydrochloric acid and poured down the drain.

58

Stars and Stripes Forever

I Hear America Singing

I hear America singing, the varied carols I hear,
Those of mechanics, each one singing his
as it should be blithe and strong,
The carpenter singing his as he measures his plank or beam,
The mason singing his as he makes ready for work, or leaves off work,
The boatman singing what belongs to him in his boat,
the deckhand singing on the steamboat deck,
The shoemaker singing as he sits on his bench,
the hatter singing as he stands,
The wood-cutter's song, the ploughboy's on his way in the morning,
or at noon intermission or at sundown,
The delicious singing of the mother, or of the young wife at work,
or of the girl sewing or washing,
Each singing what belongs to him or her and to none else,
The day what belongs to the day –
at night the party of young fellows, robust, friendly,
Singing with open mouths their strong melodious songs.

Walt Whitman, "Leaves of Grass" (1860)

Materials and Chemicals

Chromatographic paper 45 × 55 cm, 7 strips of paper 55 × 3.5 cm, chromatographic paper 12 × 12 cm, an equally large piece of transparent film, small white stars cut out of cardboard or paper, adhesive, atomiser or sprinkler, 2-L beaker, 0.1 mol/L sodium hydroxide solution, 0.1 mol/L hydrochloric acid, 0.1 % aqueous congo red solution, safety glasses, protective gloves.

Experimental Procedure

The seven strips of paper are dipped into the 0.1 mol/L NaOH solution, dried on a layer of paper and sprayed with the 0.1 % congo red solution. The square piece of paper is treated in the same way using the 0.1 mol/L HCl solution. The strips, which are now red, and the blue square are placed on the 45 × 55 cm

Congo red as an acid-base indicator

piece of chromatographic paper in such a way as to produce the United States national flag. The transparent film with the small white stars (50 if possible) is now glued to the blue square in the top left-hand corner: the star-spangled banner is finished.[1]

Explanation

Congo red is an acid-base indicator which can be used to cover the pH range 3.0 to 5.2; it is blue in acid solution and red in alkaline solution.

Reference

1 P. D. Dewar, *J. Chem. Educ.*, **1992**, *69*, 572.

59

Ion-Exchange Resins

The first observations related to the base-exchange capacity of soil were handed down by Thompson. *In 1905* Gans *prepared the first artificial ion exchanger from water glass and an aluminum salt solution.* Folin *and* Bell *used it for the determination of ammonia in urine; this was the first analytical application of an ion exchanger. The ion-exchange technique experienced an unanticipated impetus when* Adams *and* Holmes *prepared the first totally synthetic ion exchanger from sulfonated phenol and formaldehyde, and a high point when* Spedding *separated rare earths into fractions of high purity with the aid of an ion exchanger. Production was initiated at the Farbenfabrik Wolfen in 1936 on the basis of phenol resins, and the first Wofatites came onto the market in 1938.*

S. Neufeldt, "Chronologie Chemie" (1987)

Apparatus

Ion exchanger column, flat cell made of plexiglass (see colored figure), large funnel with 3 outlets, glass stopcock, two 600-mL beakers, stand, clamps, bosses, pH paper, magnetic stirrer, spatula, safety glasses, protective gloves.

Chemicals

Dowex 2 × 8 strongly basic anion exchange resin, 0.5 mol/L NaOH solution, 0.25 mol/L HCl solution, phenolphthalein, distilled water, 1 mol/L NaOH solution.

Preliminary Preparations

Several spoonfuls of the resin are allowed to swell for a short time in distilled water, made alkaline with 1 mol/L NaOH and about 1 mL of a 1 % ethanolic phenolphthalein solution added; a red coloration is observed. After stirring for about 30 minutes the resin is treated with dilute hydrochloric acid and then washed until neutral. The resin thus treated has now the same color as the remaining untreated resin to be used in the experiment.

Untreated anion exchanger resin Dowex 2 × 8 (also previously swollen in water) is now packed into the various chambers of the cell. Small amounts of

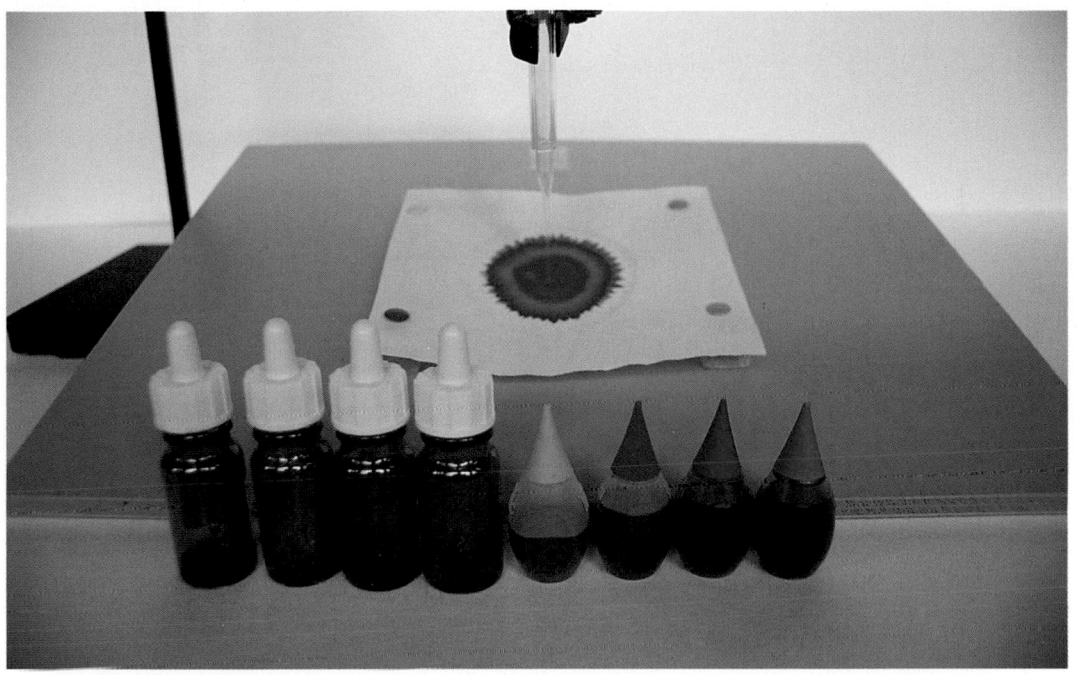

1: Experimental setup for the development of patterns according to *Runge*

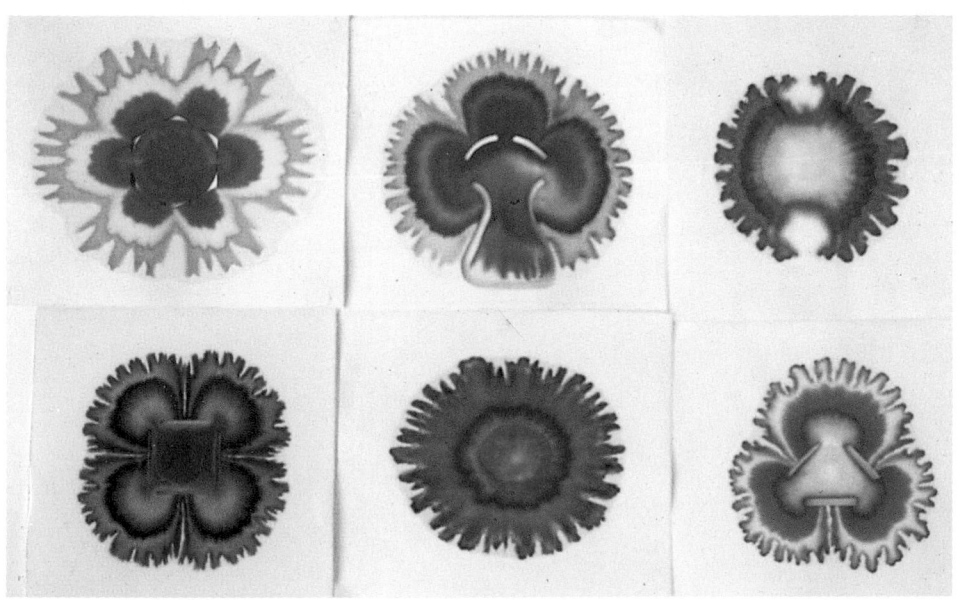

2: Patterns according to *Runge* in different tints

3: "Golden rain" of lead iodide crystals (PbI$_2$)

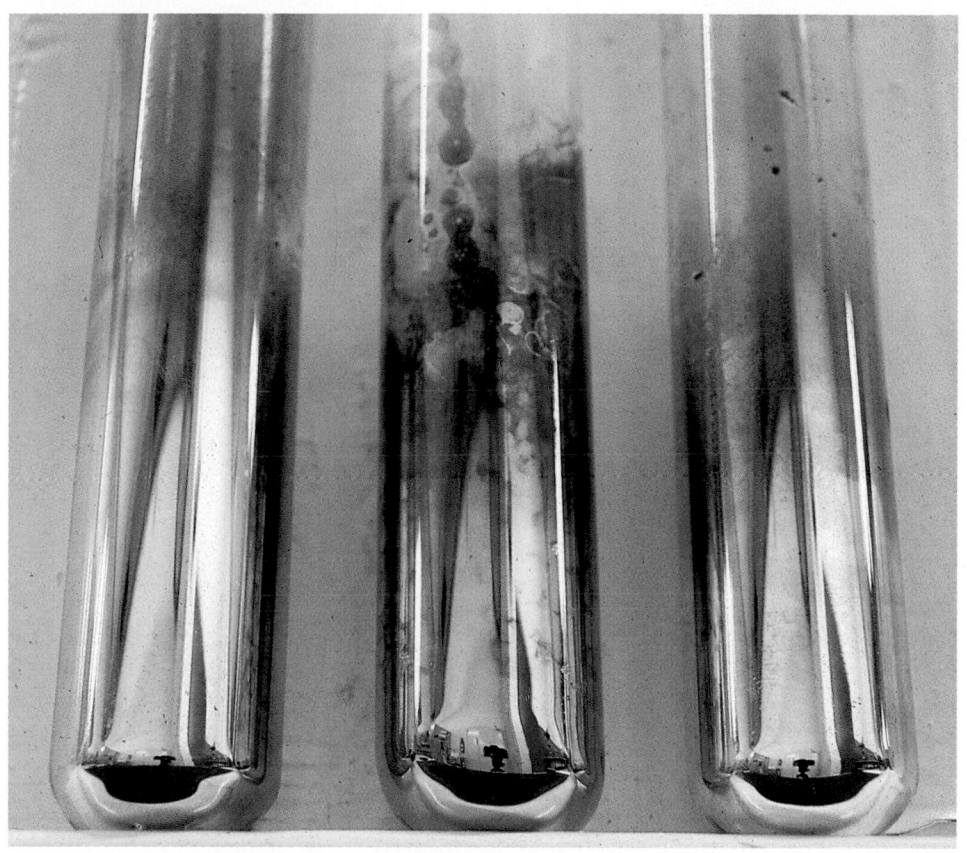

4: A comparison of copper mirrors

5: A chemical garden. Vegetation created by a combination of inorganic salts

6: Fireworks ignited by ice

7: Lightning under water

8: Growling gummy bear

9: Colors formed during the reduction of NH_4VO_3 under acid conditions

11: A reversible blue and gold reaction

10: Colors formed during the reduction of KMnO$_4$ in an alkaline medium

12: "Icecream sundae with fruit"

chymeia   al-kimiya   Chemie

chemia-chymeia-al-kimiya-Chemie

spectacular experiments

13: Magic writing

14: Color effects due to ligand exchange reactions

15: Five colors from one solution

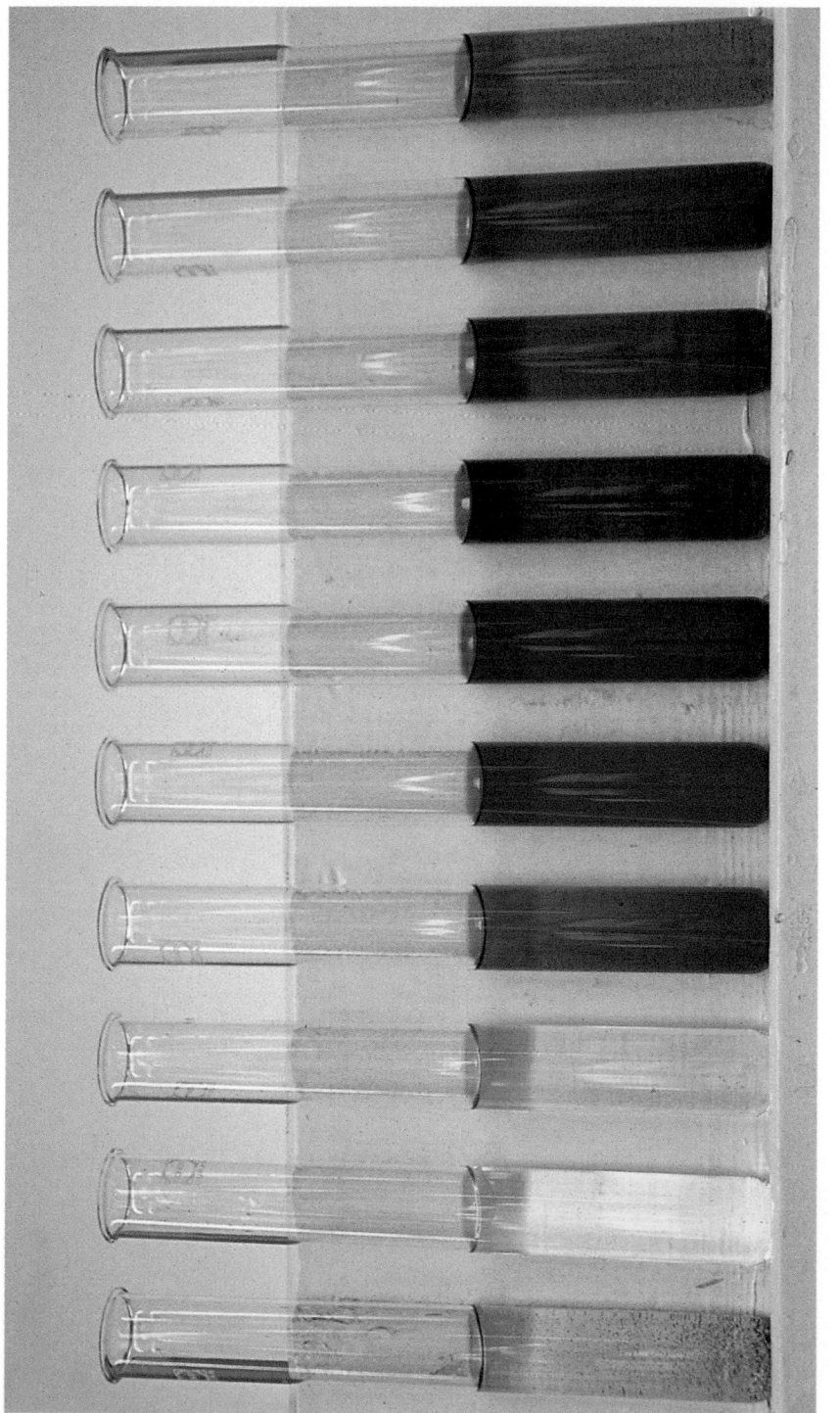

16: The influence of solvent on complex [Fe(phen)$_2$(CN)$_2$], from left to right, H$_2$O, conc. H$_2$SO$_4$, HCOOH, CH$_3$COOH, C$_2$H$_5$OH, CH$_3$OH, CHCl$_3$, CH$_2$Cl$_2$, DMFA, DMSO

17: Influence on equilibrium reactions of $CoCl_2$ in aqueous solution by adding H_2O, HCl, NO_2^- and SCN^- (from left)

18: The colors of the rainbow as formed in acid-base reactions using selected indicators

19: Color spectrum produced by an extract of red cabbage when the pH is varied

20: Colors produced by the indicators phenolphthalein, methyl red, phenol red and Yamada's universal indicator in weakly alkaline solutions. In each case two measuring cylinders are used

21: Change of the indicator color in the cylinders 2, 4, 6 and 8 on adding solid CO_2

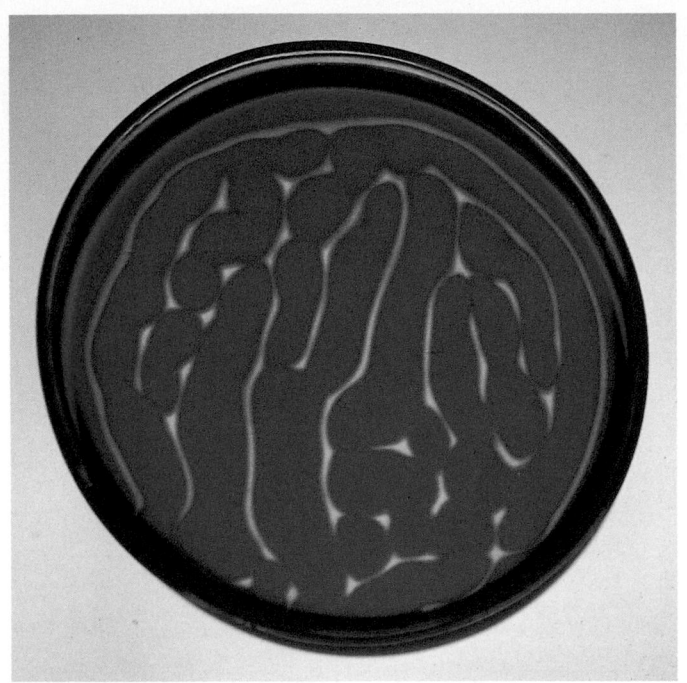

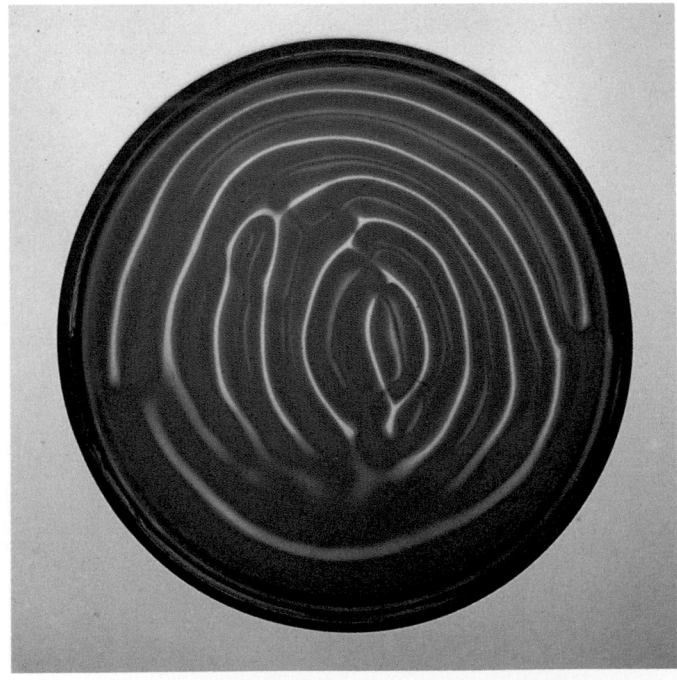

22: Formation of color patterns in a weakly alkaline solution of the indicator bromocresol green

23: The colored ion exchange resin shows which river the water comes from!

24: Chemoluminescence with luminol

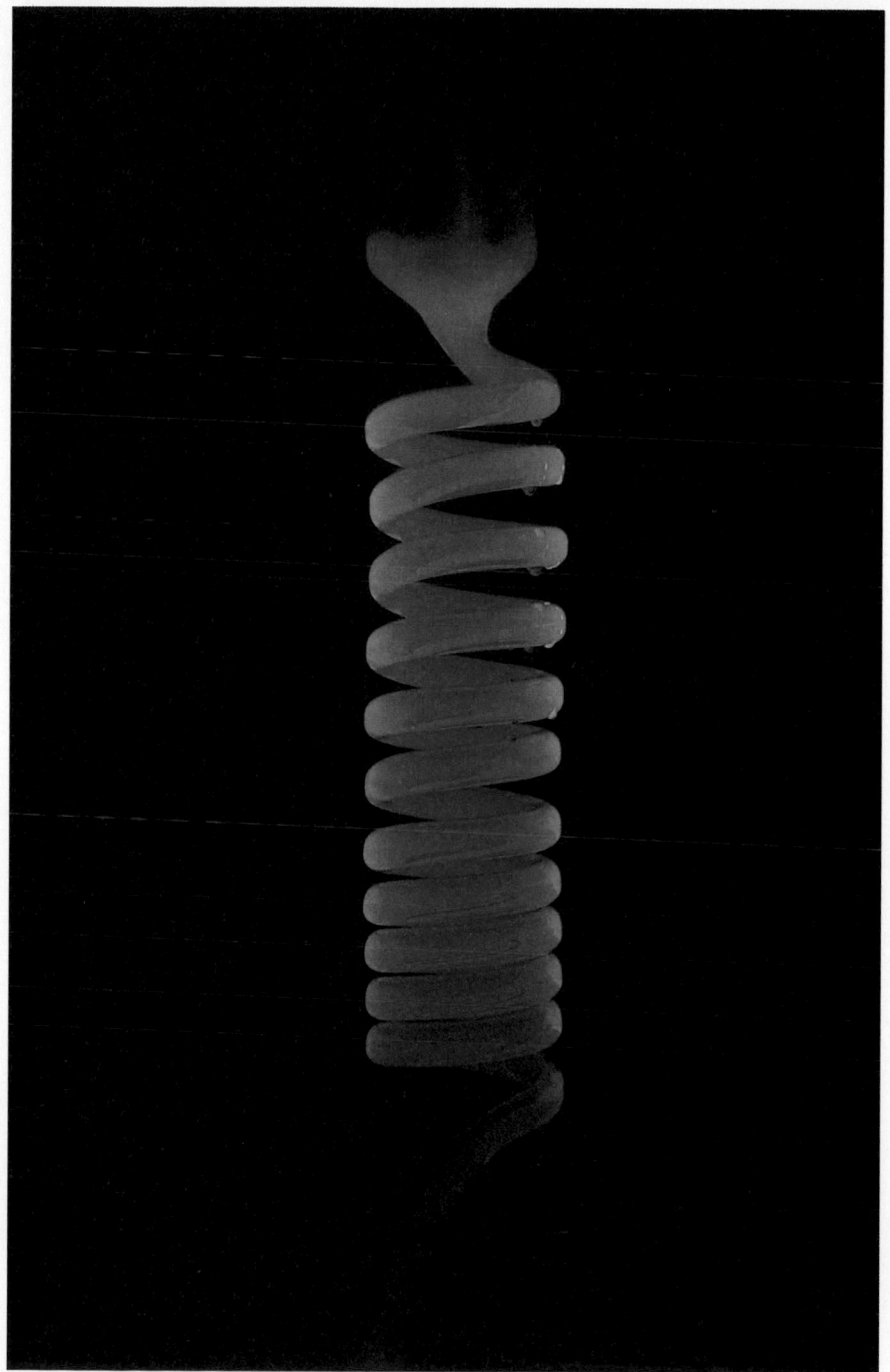

25: Chemoluminescence with luminol

26: Chemoluminescence of singlet oxygen

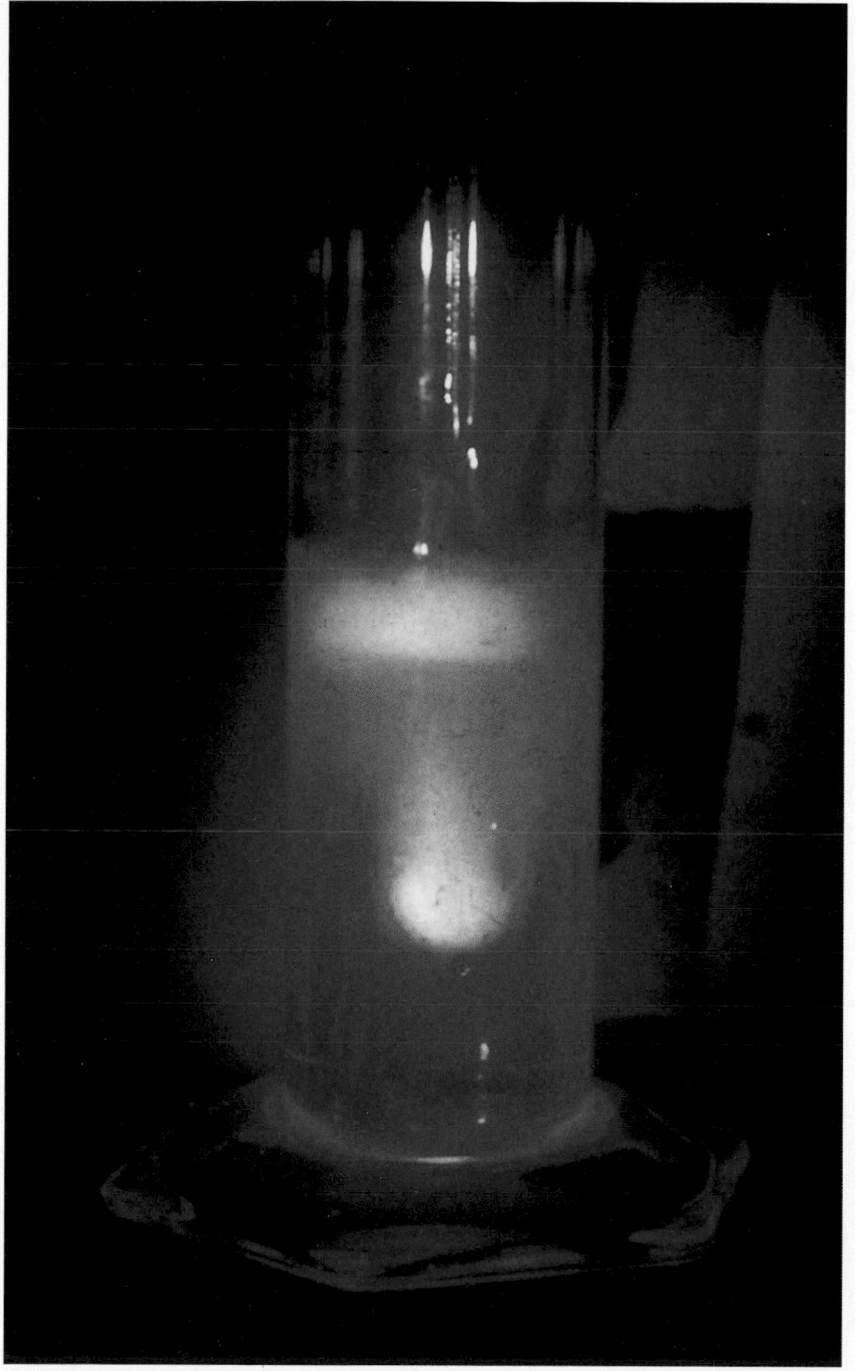

27: Chemoluminescence of singlet oxygen
(© M. Kasha, Florida State University)

28: A photochemical reaction causes the gel which has been exposed to light to turn blue

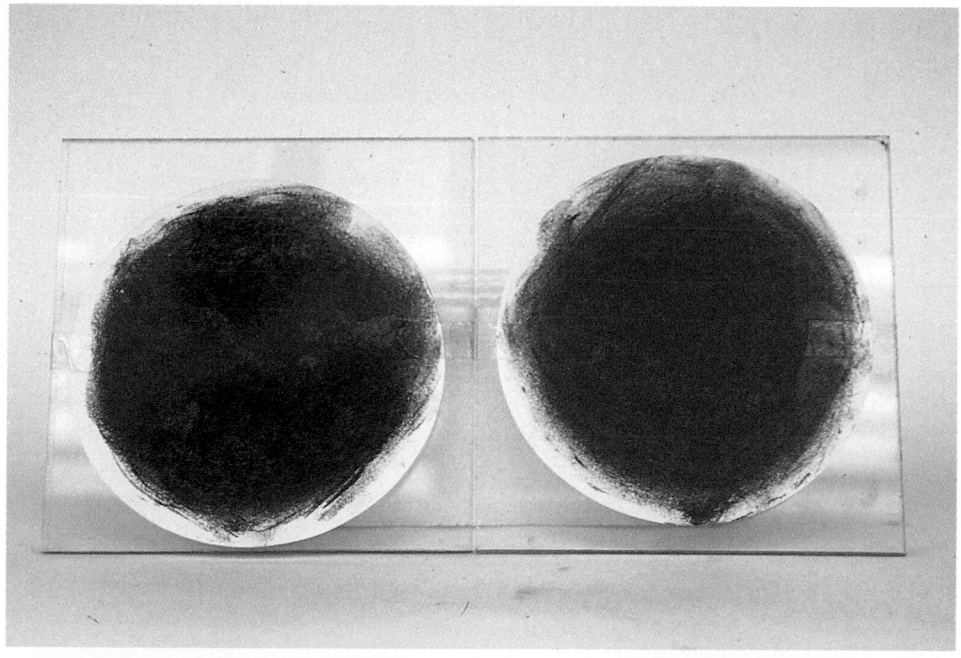

29: Thermochromism of $Ag_2[HgI_4]$ (at the top) and $Cu_2[HgI_4]$ (below)

30: The preparation of "beer"

31: Preparation of "Coca-Cola"

32: Combustion with emission of sparks

33: A burning gel

34: Flame due to burning methyl borate

35: Dissolving polystyrene foam

36: Orthorhombic sulfur crystals (α-cyclo-s₈)

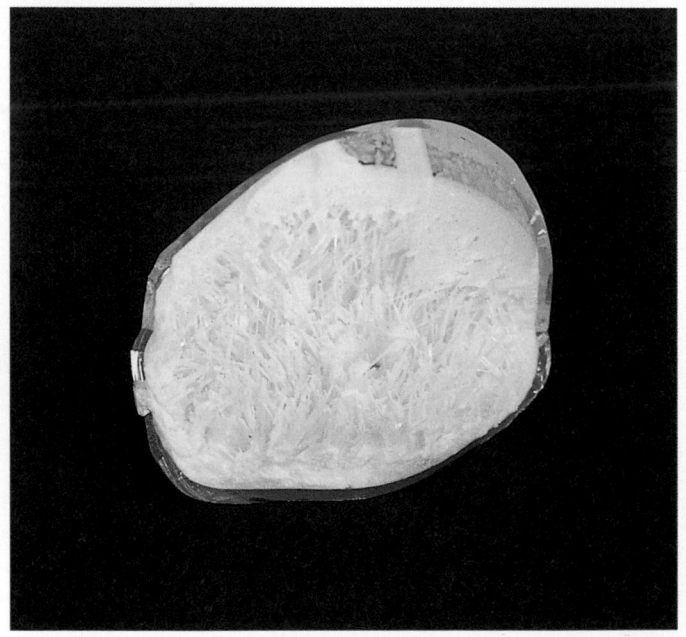

37: Monoclinic sulfur (β-cyclo-S₈)

the resin in the form of letters, numbers or other patterns are now removed with a small spatula and carefully replaced with the treated resin. The cell is closed and attached vertically to the stand. A stopcock is attached at the lower end and the funnel at the upper end (funnel height 200 mm, diameter 45 × 50 mm, volume 300–500 mL); elution is then carried out using 200 mL of the 0.25 mol/L hydrochloric acid solution.

Before carrying out the experiment the resin is finally washed with water until it is neutral.[1]

Experimental Procedure

A: about 500 mL of fresh tap water are poured into the funnel. The "water" flows through the cell within about 10 minutes. The vessel into which the tap water is filled contains (invisible for the audience) about 10 mL of 0.5 mol/L sodium hydroxide solution, so that the water is made alkaline. The treated resin (the letters etc.) turns red because of the alkaline reaction (see colored figure).

B: about 300 mL of the 0.5 mol/L sodium hydroxide is introduced onto the column from a beaker. After 5 minutes the letters appear in red. The pretreated resin contains phenolphthalein, but not the remaining resin.[1]

Waste Disposal

The polystyrene resin can be regenerated by acid. It can be disposed of via the container used for collecting non-halogenated organic solid waste.

Reference

1 D. C. Sherrington, J. Carruthers, *J. Chem.Educ.*, **1986**, *63*, 1090.

A Three-Layer Liquid

One takes a tall glass capable of holding about half an ounce or a bit more of the various liquids to be employed, and fills a quarter of it with pure iron fillings, or with mercury. Onto this one pours an equal amount of a completely saturated salt-of-tartar solution, onto this an equivalent portion of alcohol, and the remainder one fills with mineral oil or turpentine oil. These liquids remain in place, one above the other, in the order specified, because they are of different weights. Even if one shakes them together they return to their previous arrangement as soon as calm is restored in the vessel.[1]

This experiment was often used by apothecaries to obtain particularly mysterious fillings for show bottles. Such recipes are the original form of the cocktail, which was invented at the end of the 18th century and takes its name from the bright colors of the liquids, which resemble the tail feathers of a cock. Here is a modern cocktail recipe:

Brandy Cobbler

Ice cubes, 1 teaspoon raspberry syrup, 1 dessertspoon maraschino, 1 dessertspoon curaçao, 4 cl brandy, half a peach, 4 cocktail cherries, 1 pineapple cube, mineral water.

The glass is filled half full with the crushed ice cubes. Raspberry syrup, maraschino, curaçao and brandy are added, the mixture is stirred briefly and topped up with mineral water.

Naturally the chemist must show what he can do, and the following three-layer liquid is the result.

Apparatus

1-L measuring cylinder, 500-mL Erlenmeyer flask, stopper, safety glasses, protective gloves.

Chemicals

Potassium carbonate, copper sulfate, methanol, *m*- or *p*-xylene, distilled water, Sudan III dye, paraffin.

Experimental Procedure

Equal volumes of methanol and water are poured into the Erlenmeyer flask; the mixture is treated with solid K_2CO_3 until saturated (precipitate!).

Two layers are formed on mixing. A few crystals of $CuSO_4 \cdot 5 H_2O$ are now added and the mixture stirred well. The flask is closed with a stopper and left to stand for several hours. The supernatant liquid is decanted off, the solution filtered and placed in the measuring cylinder. A few mL of xylene (previously been colored with Sudan III dye) are added. This deep red liquid forms the upper phase; below it is the colorless methanol phase and at the bottom the aqueous phase which is colored blue by the copper sulfate. The measuring cylinder is closed with a cork stopper and sealed with paraffin. The mixture remains stable for several years. Prior to each demonstration the liquid is shaken thoroughly; the cylinder is placed in front of a white background and the red, white and blue cocktail effect presented.

Explanation

A multiphase system is formed because of the different densities and the limited miscibility of the liquids. The differing solubilities of suitable dyes or colored salts in the various solvents make the formation of layers separated by phase boundaries clearly visible.

Waste Disposal

Disposal is in principle not necessary, as the mixture can be kept for years. If it is to be disposed of, the upper organic layer should be poured into the container used for collecting non-halogenated organic solvents. The remainder can be poured down the drain.

Reference

1 O. Krätz, *Historisch-chemische Versuche*, Aulis-Verlag, Köln, **1987**, 45.

Some Very Different Cocktails

Olympic Cocktail

Crushed ice is placed in the cocktail mixer and 2 cl curaçao added, followed by 2 cl orange juice and 4 cl brandy. The mixture is shaken thoroughly and poured through a sieve; a beautifully colored drink is obtained.

Spicy Cocktail

3/8 L tomato juice, 1 dessertspoonful of brandy and the juice of half a small lemon are mixed with a teaspoonful of Worcester sauce and one of pepper sauce. This mixture is divided between 4 glasses and small cubes cut from a ripe pear (or small slices of banana) are added. Four teaspoonfuls of horse-radish are stirred with a little sour cream and a teaspoonful of this liquid added to each glass.

To all those who would like to try out *these* cocktails we say "Cheers!"

Safety Measures

Chemicals should *never* be stored in containers used for food!
   To be on the safe side the experiment is carried out in a glass tube about 50 cm long and 4 cm in diameter which can be closed with a stopper.

Apparatus

Glass tube 50 cm long with stopper, stand, clamp, fixing device, protective gloves, safety glasses.

Chemicals

0.5 mol/L $CuSO_4$ solution, castor oil, paraffin oil, crystals of iodine.

Experimental Procedure

150 mL of each of the following liquids are poured one after the other into the long glass tube: aqueous copper sulfate solution, castor oil and paraffin oil

colored by a few crystals of iodine; the tube is stoppered. It it is turned upside down, the CuSO₄ solution flows through the oily phases like lava, and after a while the three mixtures colored light blue, yellow and violet are clearly separated.

Explanation

Iodine dissolves very readily in the non-polar solvent paraffin and its violet color is seen in the uppermost phase. The middle layer consists of castor oil, which is naturally colored light yellow; as an ester of ricinoleic acid it can also dissolve small amounts of iodine with the formation of a yellow color. The blue copper sulfate solution forms the lower layer because of its higher density.

Waste Disposal

The two upper layers are disposed of via the container used for halogenated organic solvents, while the blue copper sulfate solution is transferred to the container used for heavy metal wastes.

To finish with, here is a "cocktail" which was often used to good effect in the 18th and 19th centuries as a display in apothecaries' shop windows.[1]

1. The bottom layer: trichloromethane colored green by maceration of grass
2. A layer of white glycerine
3. Castor oil colored red by alkanet extract
4. Spirit of wine of specific weight 0.935
5. Yellow cod liver oil
6. Spirit of wine colored by aniline blue as the top layer.

Reference

1 Cited by O. Krätz, *Historisch-chemische Versuche*, Aulis-Verlag, Köln, **1987,** 46.

62

The Nernst Distribution Law

Walter Nernst (1864–1941) is one of the most famous persons in science. In 1920 he received the Nobel Prize for his contributions to physical chemistry and his pioneering achievements in chemistry and physics. Here we shall concern ourselves only with a publication which appeared while he was working in Göttingen which combines fundamental physical understanding with practical chemical experience, entitled *"Verteilung eines Stoffes zwischen zwei Lösungsmitteln und zwischen Lösungsmittel und Dampfraum"* ("Distribution of a substance between two solvents and between solvent and vapor phase"). This appeared in 1891 in the *Zeitschrift für Physikalische Chemie* and was the first publication concerned with the Distribution Law. *Nernst* states that:

> *When a dissolved substance is distributed among different almost solphophobic liquids (e.g. by shaking water and carbon disulfide with iodine) immediately a second theorem results out of the above mentioned one. In a state of equilibrium with a given temperature the concentration ratio of the dissolved substance does not depend on the quantity of the latter, in other words, the dissolved substance has to have a constant temperature coefficient if in both solvents it meets the same molecular size.*

Nernst's acumen is demonstrated by his brilliant application of Henry's and Dalton's Laws, which deal with gas-liquid equilibria, to systems in which the substances undergoing distribution can occupy all possible aggregation states; chemical experience results in the limitation (often forgotten) of the distribution law to molecules of the same size (and in addition of the same type). Here we shall demonstrate an experiment on the distribution law which fulfils the conditions specified by *Nernst*.

Apparatus

500-mL Erlenmeyer flask, 200-mL Erlenmeyer flask, stand with fixing device, thermostatted water bath, mechanical stirrer, pipettes, Pasteur pipettes, stock bottles with stoppers, safety glasses, protective gloves.

Betain dye (BD)

Chemicals

1-Pentanol, distilled water, betain dye (see figure), triethylamine.

Experimental Procedure

50 mL of 1-pentanol and 50 mL of water are stirred for an hour at 25 °C in the 500-mL Erlenmeyer flask to achieve mutual saturation. The lower phase consists of water saturated with a very small amount of 1-pentanol, while a considerably larger amount of water (up to 30 %) is present in the alcoholic upper phase. The two liquids are carefully removed from the flask by means of pipettes and transferred to the thermostatted stock bottles. 10 mL of each of the solvents are now placed in the smaller Erlenmeyer flask; one of the samples contains about 10^{-5} mol/L of the betain dye in its basic form (obtained by adding triethylamine at about pH 10 in the aqueous phase). After stirring for 5 minutes and standing for a further 5 minutes a clearly visible phase separation occurs between the lower red aqueous phase and the upper violet alcoholic phase. If chlorinated hydrocarbons such as trichloromethane or dichloromethane are used instead of 1-pentanol, the lower organic phase is blue. The change in concentration occurring during the distribution can easily be followed quantitatively by spectrophotometric comparison measurements.

Explanation

In the aqueous phase the betain dye has the same structure as in 1-pentanol (or in chlorinated hydrocarbons), as that *Nernst's* preconditions are fulfilled. This is not the case in a number of traditional experiments involving the distribution law, such as iodine in the system water/tetrachloromethane: here iodine exists in the non-polar organic phase as iodine molecules, but in the aqueous phase as the I_3^- ion, so that comparisons are very limited. In our example the distribution coefficient K_c at the temperature of 25 °C has an average value of 4.53:

$$\frac{^cBD^I}{^cBD^{II}} = K_c = 4.53$$

I: 1-pentanol
II: water
BD: betain dye

A detailed description of this experiment and other comparable experiments for demonstrating the Nernst distribution law have recently been published by Elias et al.[2]

Waste Disposal

The solutions are collected in the container used for halogenated solvents.

References

1 W. Nernst, *Z. phys. Chem.*, **1891**, *8*, 110.
2 H. Elias, S. Lorenz, G. Winnen, *Chemie in unserer Zeit*, **1992**, *26*, 70.

63

Separation of Leaf Pigments by Column Chromatography

In the class of people who are interested in facts and ideas, we have, of course, most scientists, and also a good number of nonscientists who think along the same lines even though they don't have scientific training. In the other class – those interested in words – we have some scientists and some philosophers, and many nonscientists. I remember reading a book on philosophy in which the author went on, page after page, on the question: If there is a leaf on a tree and you see that it is green in the springtime and red in fall, is that the same leaf or is it a different leaf? Is the essence of leafness still in it? Words, words, words, but "chlorophyll" and "xanthophyll" – which are sensible in this connection of what has happened to that leaf – just don't appear at all.

Linus Pauling

Safety Measures

Toluene is toxic. The experiment must be carried out in a well-ventilated hood.

Apparatus

Glass tube about 18 cm long and 1 cm in diameter, thick glass rod to fit the tube, thin glass rod, stand with fixing device, dropping funnel, two suction bottles which can be attached to the water pump, glass frit, perforated plate to fit the glass tube, separation funnel, 250-mL Erlenmeyer flask, stopper with a hole for attaching the funnel to the glass tube, safety glasses, protective gloves.

Materials

4 fresh spinach leaves, anhydrous calcium carbonate, icing sugar, neutral activated alumina, anhydrous sodium sulfate, toluene, methanol, light petroleum (70 °C), diethyl ether, cotton wool.

Experimental Procedure

The spinach leaves (or other fresh leaves) are placed in a mixture of 45 mL of light petroleum, 5 mL of toluene and 15 mL of methanol in the Erlenmeyer flask. After standing for an hour, the residue, which is almost white, is filtered off and washed with the same solvent mixture. Methanol is removed by careful treatment with water in the separating funnel (do not shake!) and the remaining mixture is dried over Na_2SO_4. A glass tube about 18 cm long and 1 cm in diameter with a bulb at its lower end is used for the separation. A well-fitting perforated plate is now introduced and covered with a small plug of cotton wool. Aluminium oxide is introduced in small portions into the tube until the layer is 2 cm thick. The column is packed by pressing gently with the thick glass rod, which just fits the tube. A 4 cm layer of calcium carbonate (previously dried at 150 °C) is now introduced, followed by a 6 cm layer of icing sugar which has been sieved finely. A slight vacuum, which should be kept constant, is now applied to the tube while light petroleum is allowed to flow in from the dropping funnel; the green solution is then passed through the tube without letting the adsorbents come in contact with air. Several colored zones are formed; the upper yellow-green layer contains chlorophyll b and the blue-green layer below it chlorophyll a. A yellow zone containing xanthophyll is below this. The yellow carotine is absorbed by the Al_2O_3, forming a narrow zone.

When the dye solution has almost completely passed through the column the "chromatogram" is "developed", i.e. the zones are separated by washing with a mixture of light petroleum and toluene (4:1). If necessary more light petroleum can be added to keep the zones from spreading out. The column is then washed with light petroleum, sucked dry and the adsorption column pushed out of the tube with a suitable glass rod. The color zones can then be cut apart. The dyes are dissolved in ether containing a little methanol and their absorption spectra determined.[1] The separation of the dyes by paper chromatography as suggested by *Bukatsch et al.*[2] is also very well suited for studying the complex nature of the photosynthetically active leaf dyes.

Explanation

The blue-green chlorophyll a and the yellow-green chlorophyll b contain a porphyrin system with various side chains and complexed magnesium (II); chlorophyll b bears a -CHO group in ring II, chlorophyll a a methyl group. The separation of these two components was first carried out in 1913 by *R. Willstätter* (1872–1942), and in 1960 *R. B. Woodward* (1917–1979) and his co-workers were able to report the total synthesis of chlorophyll. Carotine and its relatives the carotinoids are symmetrically constructed hydrocarbons containing conjugated double bonds; because of their orange-yellow color and their low solubility in water they are also referred to as lipochromes. Their oxidised derivatives include the leaf xantophyll lutein.

Waste Disposal

The organic liquids are transferred to the corresponding container. All other substances are washed down the drain with copious amounts of water.

References

1 L. Gattermann, H. Wieland, *Die Praxis des Organischen Chemikers*, 34. Aufl., Walter de Gruyter & Co., Berlin, **1952,** 359.
2 F. Bukatsch, O. P. Krätz, G. Probeck, R. J. Schwankner, *So interessant ist Chemie,* Aulis-Verlag, Köln, **1987,** 8.

Separation of the Colored Inks from Felt-Tip Pens

Some years ago *E. Halpapp* and *H. E. Hauck* from the chromatography research laboratory at E. Merck in Darmstadt developed a novel technique for the creation of flow pictures with various combinations of colors and shapes, which they called CHROM ART. CHROM ART found its origin in a problem involving pure analytical science. Mixtures of colorings are applied, generally over a large area, to TLC plates coated with silica gel either prior to or during the development of the chromatograms. During this development pictures are formed by the interactions between the system of pores in the silica gel and the dyestuff molecules; these pictures can produce an amazing variety of aesthetically appealing color effects which are determined by the creativity of the "artist". This "painting" technique forms an inseparable part of the scientific and artistic activity of *H. Halpaap*.[1]

Even a simple thin layer chromatographic separation of the various colors in felt-tip pens shows the power of this analytical procedure.

Safety Measures

The experiment must be carried out in a well-ventilated hood, as solvents may be set free.

Apparatus

Glass vessel (20 × 10 × 5 cm) with lid, micropipettes, 1, 2 or 5 μl, safety glasses, protective gloves.

Chemicals

Aluminum TLC foil coated with silica gel 60, toluene, acetone, methanol, ethanol, several different colored felt-tip pens.

Experimental Procedure

Different colored felt-tip pens are extracted with either acetone, methanol or ethanol; the starting spots are applied to the TLC foil about 15 mm from its lower end with a micropipette. The glass vessel contains a few mL of the mobile phase, in this case a mixture of toluene, acetone and methanol in the ratio 7 : 3 : 2. The TLC plate is now placed vertically in the solution. The solvent front rises slowly and evenly, the colored origins undergoing separation into the various components of the dye. The experiment takes 30 to 40 minutes.

Explanation

The rate of migration of the color components is determined by their molecular mass and the strength of the interactions between the components and the solvent (mixture).

Waste Disposal

The remainder of the mobile phase can be stored for re-use.

Reference

1 E. Halpaap, H. E. Hauck: German Patent Application P 3204094.6 (6. 2. 1982); *Merck-Spektrum*, **1992**, *9*, 25.

65

Chemoluminescence

He who wishes to learn and to discover the secrets of Nature, let him begin with this ...

Giambattista della Porta (around 1589)

Hennig Brand is considered to be the discoverer of phosphorus. The glow of white phosphorus in the dark is due to its oxidation to phosphorus trioxide, which then undergoes oxidation to phosphorus pentoxide with the emission of light. In 1669 *Brand* concentrated urea to dryness and heated the residue strongly. He obtained a product which glowed in the dark, because the phosphate $NaNH_4HPO_4$ present in urine reacted as follows with the carbon formed during calcination:

$$NaNH_4HPO_4 \rightarrow NaPO_3 + NH_3 + H_2O$$

$$4\ NaPO_3 + 5\ C \rightarrow 2\ Na_2CO_3 + 3\ CO_2 + P_4$$

Duke *Johann Friedrich* of Brunswick-Lüneburg-Hanover wanted *Brand* to be his personal physician and alchemist at the court in Hanover. He hoped that *Brand's* universal medicine would allow him to drive out the devil and find large amounts of gold and silver in the Harz mountains.

In 1676 *Gottfried Wilhelm Leibniz* was employed in Hanover as librarian and historian; his title was Hofrat. The Duke ordered him to convince *Brand* to join the Hanoverian court. *Brand* was invited to Hanover, and during the negotiations on compensation for loss of income during his stay in Hanover he talked of having to feed eight mouths. In a letter of 26th November 1678 he estimated his loss of income due to the journey to Hanover to be 300 reichsthaler. Of course he was just boasting in order to try to squeeze more money out of the Hanoverian court. The Duke and *Leibniz* knew that *Brand* was deeply in debt in Hamburg, although his wife *Margarethe* had brought a large fortune into the marriage.[1]

Safety Measures

The toxicity of luminol is not known exactly. Sensibilisation due to breathing in luminol dust and to skin contact is possible. Potassium hexacyanoferrate(III) is toxic. Skin contact must be avoided. 30 % H_2O_2 solution is a strong oxidising agent. Skin contact can lead to serious burns. Protective gloves and safety glasses should be worn. H_2O_2 can undergo spontaneous decomposition in the presence of metals or organic compounds. Contact with NaOH can cause severe skin damage.

Apparatus

20-mL round-bottomed flask, funnel, two 600-mL beakers, two 400-mL beakers, two 50-mL measuring cylinders, protective gloves, safety glasses.

Chemicals

Luminol(5-amino-1,2,3,4-tetrahydrophthalazine-1,4-dione),   10 %   NaOH solution, potassium hexacyanoferrate(III) $K_3[Fe(CN)_6]$, 30 % H_2O_2 solution, distilled water.

Solution A

1 g luminol and 50 mL 10 % NaOH solution in 450 mL H_2O.

Solution B

500 mL 3 % $K_3[Fe(CN)_6]$ solution (15 g $K_3[Fe(CN)_6]$ in 485 mL H_2O).
   The following solutions must be prepared prior to the demonstration:

Solution C: 50 mL of solution A in 350 mL H_2O
Solution D: 50 mL of solution B in 350 mL H_2O and 3 mL 30 % H_2O_2

Experimental Procedure

A few crystals of $K_3[Fe(CN)_6]$ are placed in the round-bottomed flask. The solutions C and D are added simultaneously via the funnel. A light blue glow is observed, which can be regenerated several times by adding small amounts of the NaOH solution (color figures 24, 25). Instead of pouring the solutions into a flask they can be allowed to flow through a large glass spiral into a large beaker containing a few crystals (about 0.1 g) of $K_3[Fe(CN)_6]$.

Explanation

Luminol (I) exhibits chemoluminescence in alkaline solution. The light intensity can be increased by catalysts such as $K_3[Fe(CN)_6]$, which however also increase the rate of decay of the light emission. During the reaction the luminol is converted into the disodium salt of 3-aminophthalic acid (II).

Oxidation of luminol

Waste Disposal

The solution is concentrated on a hotplate to a volume of 100 mL. This must be done in a well-ventilated hood! The residue is transferred to the container used for heavy metal waste.

Reference

1 F. Krafft, *Phosphorus. From elemental light to chemical element, Angew. Chem.*, **1969**, *81, 634; Angew. Chem. Int. Ed. Engl.* **1969**, *8*, 660.

66

Two-Color Chemoluminescence

The world of chemical events resembles a stage on which an unbroken succession of scenes is played out. The cast consists of the elements. Each is assigned its unique role, be it that of walk-on or principal character.

Clemens Winkler

In the kingdom of sagas and fairy tales one imagines tiny mushrooms in the rain forest of the equatorial zone that emit a magical green glow, thereby conjuring up a starry firmament under the foliage. But these mushrooms are not the only forms of life with the ability to produce cold light. This phenomenon, known as bioluminescence, is based on a relatively complicated chemical oxidation process in which particular enzymes unique to each species, so-called luciferases, act as catalysts and consume a luminescent material such as luciferin. This produced either a faint shimmer or in some cases even a bright flash. Certain beetles living in South America, the cucujos, emit such a bright greenish light that the natives use the insects as reading lamps. In the abyss, at the bottom of the oceans, there are numerous types of fish that bear white, yellowish, green, blue, or reddish points of light in streaks along their bodies. These shining markings often permit members of a given species to recognize each other, or serve as spotlights for illuminating their hunting grounds. Deep sea representatives of the angler or monkfish family possess a unilobed skin appendage in the vicinity of the head that serves as illuminated bait for attracting living prey. Certain cephalopods are in a position to emit enveloping clouds of sparks, and they then avail themselves of this luminous paint as camouflage for an escape. Deep sea bass carry on their skin an envelope filled with shimmering bacteria whose illumination they can dim by throttling the flow of blood and hence the oxygen supply. The transparent purple lamp jellyfish found in the Atlantic and the Mediterranean is particularly prone to blink when it is disturbed mechanically. The luminous organisms of the deep seas, such as the flitting shrimps, have so far been studied relatively little; the shallow waters of the Indo-Malaysian Sea are home to a vertebrate that is capable of controlling the light it emits. The lantern fish directs its searchlight into the waters ahead of it like a headlamp. By rotating inward a muscle in the luminescent organ it can dim the light or direct it in a single spot onto a target.

... Fireflies and glowworms are a popular sight on warm June nights. Their light sources are derivatives of lipids. These tiny searching or flying lanterns are thus able to generate a cold, yellow-green luminescence with a light yield of nearly 100 % and no concurrent increase in temperature.

„Meisterwerke der Schöpfung,"
Pro Terra Bücher Verlagsgesellschaft mbH, Munich

Safety Measures

Safety glasses and rubber gloves must be worn during the preparations for carrying out these experiments. Pyrogallol is poisonous and can be absorbed via the skin. 30 % hydrogen peroxide is a strong oxidising agent; it must not be allowed to come in contact with organic substances or oxides and metals with clean surfaces, since these cause its rapid decomposition. Contact with NaOH can lead to severe skin damage. Formaldehyde vapor irritates the nasal mucous membrane. Experiments with rats have demonstrated that these developed cancer of the nose when they were exposed to formaldehyde vapors at a concentration of 15 ppm for 18 months.

Apparatus

250-mL beaker, 1-L beaker, large glass trough, 100-mL measuring cylinder, magnetic stirrer, rubber gloves, safety glasses.

Chemicals

NaOH, luminol, K_2CO_3, pyrogallol, 30 % H_2O_2 solution, distilled water, dilute hydrochloric acid.

Experimental Procedure

40 mL of distilled water are poured into the 250-mL beaker which is standing on the magnetic stirrer. 0.8 g of NaOH pellets are dissolved with stirring, followed by 0.005 g of luminol, 25.0 g of K_2CO_3 and 1.0 g of pyrogallol. When dissolution is complete (the solution is brown) 10 mL of a 40 % formaldehyde solution is added. This solution is now poured into a 1-L beaker which is standing in the glass trough. The room is darkened and 40 mL 30 % H_2O_2 added to the solution (stirring is not necessary). For about 10 seconds the solution glows a dull red; this color is then replaced by a bright blue glow, which again persists for about 10 seconds. The solution becomes hot and effervesces strongly.

Explanation

The chemoluminescence of luminol has been explained in Experiment 65. The red coloration is due to the formation of singlet oxygen during the oxidation of pyrogallol and formaldehyde by the aqueous H_2O_2 solution.

Waste Disposal

The solution is neutralized with dilute hydrochloric acid and transferred to the container used for collecting water-soluble organic solvents.

Reference

1 B. Z. Shakhashiri, *Chemical Demonstrations. A Handbook for Teachers of Chemistry,* University of Wisconsin Press, Madison, London, **1983,** *1,* 175.

Chemoluminescence
with Oxalyl Chloride

Me thinks I see these things with parted eye,
When everything seems double.

William Shakespeare, "A Midsummer Night's Dream"

Safety Measures

This experiment must be carried out in a well-ventilated hood. Oxalyl chloride and dichloromethane are toxic. Safety glasses and gloves must be worn. The toxicity of the indicators is not known, so that they should be handled with care.

Apparatus

Five 250-mL Erlenmeyer flasks with stoppers, 150-mL measuring cylinder, 15-mL measuring cylinder, 2-mL pipette, safety glasses, protective gloves.

Chemicals

Dichloromethane, 3 % H_2O_2 solution, oxalyl chloride ($Cl_2C_2O_2$), rhodamine 6 G, rubrene, 9,10-diphenylanthracene, 13,13'-dibenzanthronyl.

Oxalyl Chloride Solution

50 mL of CH_2Cl_2 are mixed with 2 mL of oxalyl chloride in an Erlenmeyer flask. (This solution is stable for a considerable length of time if the flask is stoppered and kept in the dark).

Experimental Procedure

0.005 g of one of the indicators, 25 mL of CH_2Cl_2 and 4 mL of 3 % H_2O_2 are introduced one after the other into a 250-mL Erlenmeyer flask. 2 mL of the oxalyl chloride solution are added, the room darkened and the flask swirled. A bright glow can be seen immediately. The color, intensity, and duration of the chemoluminescence depend on the indicator used as shown in the following table

indicator	color	intensity	mean duration
rubrene	yellow	very bright	2 minutes
9,10-diphenylanthracene	blue	very bright	3 minutes
mixture of rubrene and 9,10-diphenylanthracene	yellow to blue	bright	3 minutes
rhodamine 6 G	orange	very bright	30 seconds
13,13'-dibenzanthronyl	green to yellow to red	bright	40 seconds

Waste Disposal

The solution is placed in the container used for collecting halogenated organic solvents.

Reference

1 B. Z. Shakhashiri, *Chemical Demonstrations. A Handbook for Teachers of Chemistry,* University of Wisconsin Press, Madison, London, **1983,** *1,* 153.

68

Singlet Oxygen

We see in nature not words,
but rather only the first letters of words,
and if we then wish to read, we discover
that the new so-called words
are again merely first letters of others.

Georg Christoph Lichtenberg

The bleaching action of sunlight in the presence of air had long suggested the presence of an active form of oxygen. This existence of this species, singlet oxygen ($^1O_2$) was demonstrated experimentally by *Kautsky* and *DeBruijn*. In its ground state oxygen has two unpaired electrons and reacts in a stepwise manner as a diradical ($^3O_2$), while singlet oxygen has paired valence electrons. Two electronically excited states of $^1O_2$ have been detected. These are denoted as $^1\Delta_g$ and $^1\Sigma_g^+$ and lie approximately 96 kJ and 160 kJ above the ground state of triplet oxygen. *H. Kautsky* and *H. DeBruijn* report as follows[1]:

Explanation of the Development of Photoluminescence in Fluorescing Systems as a Result of Oxygen: The Formation of Active, Diffusible Oxygen Molecules through Sensitization.

The fluorescence and phosphorescence of numerous dye molecules adsorbed at interfaces is diminished or completely extinguished by molecular oxygen. We have now elucidated the basis of this oxygen influence, which we first recognized as a fundamental consequence of a photodynamic oxidation effect induced by the illumination of fluorescent dyes; specifically, it results from the formation of a short-lived, very active, diffusible form of the oxygen molecule that arises from the transfer of activation energy from the fluorescent dye to oxygen. The amount of energy required for activation is relatively small, since even the deep red fluorescence of chlorophyll can be largely extinguished by oxygen.

Safety Measures

This experiment should only be carried out in a well-ventilated hood. Safety glasses and protective gloves must be worn. Chlorine gas is highly toxic!

Sodium hydroxide solution and H_2O_2 are very corrosive! Skin contact must be avoided at all costs.

Apparatus

Cl_2 gas cylinder or Kipp's gas generation apparatus, 500-mL wash bottle with glass frit, 250-mL and 150-mL beakers, large plastic basin for use as an ice bath, two measuring cylinders, funnel, stand, clamp and boss, PVC tubing, protective gloves, safety glasses.

Chemicals

Sodium hydroxide pellets, 30 % hydrogen peroxide solution, distilled water, ice, chlorine from gas cylinder (if this is not available, chlorine can easily be generated from $KMnO_4$ and concentrated hydrochloric acid in a Kipp's apparatus).

Experimental Procedure

Prior to the experiment the 250-mL beaker, which contains 20 g of NaOH in 140 mL of water, and the 50-mL beaker containing 30 mL of 30 % H_2O_2 solution are cooled in the ice bath. The two well-cooled solutions are poured into the wash bottle and the room darkened. A rapid stream of chlorine gas is now passed through the solution, which at once glows bright red[2] (colored figures 26, 27).

Explanation

The initial product formed when chlorine is passed through an alkaline hydrogen peroxide solution is hypochlorite:

$$Cl_2 + 2\,OH^- \rightarrow OCl^- + Cl^- + H_2O$$

The hypochlorite ion probably reacts with hydrogen peroxide to give the chloroperoxide ion, which splits off chloride to form oxygen:

$$H_2O_2 + OCl^- \rightarrow ClOO^- + H_2O$$
$$ClOO^- \rightarrow {}^1O_2 + Cl^-$$

According to the law of conservation of spin oxygen is formed in the excited singlet state rather than the triplet state, which is the ground state of molecular oxygen. Triplet oxygen can be converted into electronically excited singlet oxygen by supplying energy. The excited oxygen molecules emit red light on returning to the triplet ground state.

$$^1O_2 \rightarrow {}^3O_2 + h\upsilon \; (\lambda = 634 \; nm)$$

The chemical behavior of the two types of oxygen is also quite different.

Waste Disposal

The reaction solution is diluted with a large amount of water, neutralised with sulfuric acid and poured down the drain.

References

1 H. Kautsky, H. DeBruijn, *Naturwissenschaften,* **1931,** *19,* 1043.
2 W. Adam, W. Baader, *Singulettsauerstoff – chemische Erzeugung und Chemolumineszenz, Chemie in unserer Zeit,* **1982,** *16,* 169.

69

Generation of Singlet Oxygen
in the Presence of Dyestuffs

Fantasy is more important than knowledge.

Albert Einstein

It is possible to obtain other colors by preparing singlet oxygen in the presence of dyes (photosensibilisation).

Safety Measures

See experiment 68.

Apparatus

250-mL wash bottle, 50- and 100-mL measuring cyclinders, Kipp's apparatus or chlorine gas cylinder, protective gloves, safety glasses.

Chemicals

20 mL 30 % H_2O_2 solution, 60 mL 6 mol/L NaOH, 20 mL CH_2Cl_2, luminol, indanthrene dark blue (dibenzanthrone) or indanthrene brilliant blue (violanthrone).

Experimental Procedure

A small amount (about 0.01 g) of indanthrene dark blue is placed in a wash bottle. 20 mL of CH_2Cl_2, 60 mL of 6 mol/L NaOH and 20 mL of a 30 % H_2O_2 solution are added. The room is darkened and chlorine passed through the solution. After a short time the gas bubbles are surrounded by a beautiful carmine glow. A lovely lavender blue glow is obtained when luminol is used.

Explanation

(simplified equation):

$$O_2^* + V \rightarrow O_2 + V^*$$

$$V^* \rightarrow V + h \cdot v \ (V = dye)$$

Waste Disposal

The reaction mixture is transferred to the container used for collecting halogenated solvents (avoid skin contact!).

70

The Mitscherlich Test

But money is the sovereign spell,
I can think of nothing better in this world
Than poison and money.
Poison creeps in darkness into money and power,
I have taken it to be my companion,
And found it to be good.
Down I plunged into the kingdom of shadows of
Man, brother, father, and became at once
Respected and rich.

Adalbert von Chamisso, "The Poisoner"

They took a beer keg and partially filled it with straw of peas on which the horse had frequently urinated. Likewise dry bloodwort that had been strewn under the swine, pils seeds or the plant itself, dried cow gall, decayed and stinking tench chopped into pieces, and to this they added as well a pound of wax, a pound of unßlit and bacon, dry powder, operment and flue dust, all of this mixed together until the keg was full. When they were ready to use it they loaded it onto a cart, turned the opened end toward the enemy, and then ignited the material and pushed the cart toward the barricades or storage depots, preferably at a time when the wind was blowing toward the enemy. Then, according to their records, the smoke poured out and caused the enemys bodies to swell, and within three hours they were made unconsious, indeed a great many were killed.

The phosphorus fire bombs used in the Second World War repesent a sorry climax of the development of this principle.

Safety Measures

White phosphorus is highly toxic: 0.05 g taken orally can be lethal. White phosphorus can ignite spontaneously at room temperature when finely divided. Avoid skin contact, it causes severe burns.

Apparatus

500-mL round-bottomed flask with 29/42 ground glass joint, bulb condenser, heating mantle, stand, clamp and boss, boiling stones, tweezers, safety glasses, protective gloves.

Chemicals

White phosphorus, distilled water, 1 mol/L $CuSO_4$ solution.

Experimental Procedure

300 mL of water are placed in the flask, followed by a piece of phosphorus the size of a pea. The flask is now heated and the bulb condenser attached; no cooling water is used. As soon as the water begins to boil the room is darkened. A pale blue flame is visible at the lower end of the bulb condenser, and moves slowly upwards as the heating is continued. A strip of paper which is held in this "flame" does not catch fire.

The reaction can also be carried out in an Erlenmeyer flask to which is attached a glass tube one meter long and 0.5 cm in diameter.

A particularly striking effect is observed when a 2 % solution of white phosphorus in glacial acetic acid is treated with a 3 % hydrogen peroxide solution.

Explanation

The bluish glow emitted by white phosphorus is due to an oxidation process in which the phosphorus-containing vapor is first oxidised by air to low-valent phosphorus oxides, these in turn being converted to P_4O_{10}, a process which is accompanied by light emission.

The *Mitscherlich* test is used in forensic medicine for determining whether phosphorus is present in the stomach.

Waste Disposal

Phosphorus residues are destroyed using the copper sulfate solution. The copper phosphide formed should be transferred to the container used for heavy metal waste.

71

The Chemoluminescence
of Phosphorus

In the year 1610 the Italian painter *Guido Reni* (1575–1642) created the famous ceiling fresco *"Aurora"* in the casino of the Palazzo Rospigliosi in Rome; this depicts the triumph of the sun god, who was accompanied by the legendary Phosphorus, shown here as a torch-bearing winged youth.

When in the year 1669 the Hamburg alchemist *Hennig Brand* obtained a glowing substance by heating urine residues in a tightly closed retort he believed he had been able to separate fire in its pure form, namely phosphorus. *G. W. Leibniz* wrote as follows in his *"Historia inventionis Phosphori"* in 1710: *"At first he obtained vapors, then a sticky mass, and finally a substance of solid, granular consistency distilled over; this adhered to the walls of the still receiver like sugar and, like the whole distillate, glowed brightly in the dark."*[1]

Ten years later *Robert Boyle* confirmed *Brand's* findings, but observed that the luminescence only occurred in the presence of air, and that the vapors observed by *Brand* as well as the residue were acids. *Boyle* named the substance *"aerial noctiluca"*, and described it in his publication *"The Aerial Noctiluca or new Phaenomena and a Process of a Factitious self-shining Substance"* which appeared in 1680. The name phosphorus (from the Greek word for "bearer of light") was however soon commonly used. Just almost 100 years later *A. L. Lavoisier* showed that phosphorus was in fact an element.

Safety Measures

White phosphorus is highly toxic and ignites spontaneously at room temperature. Protective gloves and safety glasses must be worn!

Apparatus

1-L round-bottomed flask, gas burner, crucible tongs, large beaker, glass rods, safety glasses, protective gloves.

Chemicals

Castor oil, small pieces of white phosphorus, saturated milk of lime, 10 % sodium hypochlorite solution, $CuSO_4 \cdot 5\,H_2O$, glass wool.

Experimental Procedure

About 200 mL of castor oil are introduced into the round-bottomed flask, followed by a few small pieces of white phosphorus. The flask is stoppered with a plug of glass wool and shaken until the phosphorus has dissolved. The room is darkened and the mixture warmed using the gas burner; the flask is rotated carefully. If it is now shaken to and fro the oily mixture begins to phosphoresce brightly; this can be observed clearly even in a large room.

Explanation

The emission of yellow-green light during the slow oxidation of the vapor of the P_4 modification of phosphorus to P_4O_{10} is one of the oldest known examples of chemoluminescence. The species which emit light in the visible region of the spectrum are probably $(PO)_2$ and HPO; excited states of $(PO)_{1,2}$ also emit light in the UV region.[2]

Waste Disposal

The contents of the flask are shaken for a considerable time with aqueous copper sulfate solution and left to stand for several hours in the well-ventilated hood. The remaining oil (which must be free of phosphorus) is transferred to the container used for halogen-free organic solvents, while the aqueous suspension is oxidised by adding a strongly alkaline 10 % NaOCl solution dropwise with cooling and then stirred with milk of lime. The precipitate, which consists of copper phosphate and hydroxide, is disposed of with the less toxic inorganic waste, and the aqueous solution neutralised and poured down the drain.

References

1 Cited by H. W. Prinzler, *Phosphorus, Sulfur, Silicon,* **1993**, *78*, 1.
2 R. J. van Zee, A. U. Kahn, *J. Am. Chem. Soc.,* **1974**, *96*, 6805.

Chemoluminescence with Oxalic Esters

How art thou fallen from heaven,
O day star, son of the morning!
how art thou cut down to the ground,
which didst lay low the nations!
And thou saidst in thine heart,
I will ascend into heaven,
I will exalt my throne above the stars of God;
and I will sit upon the mount of congregation,
in the uttermost parts of the north:
I will ascend above the heights of the clouds;
I will be like the Most High.
Yet thou shalt be brought down to hell,
to the uttermost parts of the pit.

The Book of Isaiah 14, 12–15

In ancient Greece Venus was known as *Hesperos*, the evening star, while the morning star was called *Phosphorus*; to the Romans it was known as *Lucifer*.

Sources of artificial light are now an integral part of our everyday life. Here we shall demonstrate the reaction which form the basis of the *"light sticks"* which are available commercially as an emergency light source. The long drawn out chemoluminescence effect is caused by mixing an aromatic oxalate ester and a fluorescent dye with hydrogen peroxide in the presence of a weak base which functions as the catalyst. The reaction takes place in a polythene tube which contains a glass ampoule with the peroxide solution.

Safety Measures

30 % H_2O_2 is highly corrosive. Safety glasses and protective gloves should be worn.

Apparatus

Two 200-mL beakers, cooling bath, magnetic stirrer, reclosable polythene bottle, vacuum distillation apparatus, safety glasses, protective gloves.

Chemicals

30 % hydrogen peroxide, dimethylphthalate, *t*-butanol, sodium salicylate, fluorescent dyes:
(I)   9,10-diphenylanthracene DPA,
(II)   9,10-bis(phenylethinyl)anthracene BPEA,
(III) rhodamine B,
        bis(2,4-dinitrophenyl)oxalate DNPO (m.p. 192–194 °C),
        bis(2,4,6-trichlorophenyl)oxalate TCPO (m.p. 196–198 °C).

Experimental Procedure

Solution A

A solution of 0.01 mol/L oxalate and 0.003 mol/L (approx. 0.1 g/100 mL) of one of the dyes in dimethylphthalate is prepared. The DNPO solution must always be freshly prepared, while that of TCPO can be kept indefinitely.

Solution B

8 mL of 30 % H_2O_2 are mixed with 80 mL of dimethylphthalate in 20 mL of *t*-butanol without warming. The luminescence can be made brighter by adding 0.02 to 0.03 g of sodium salicylate.

The two solutions are filled in polythene bottles, which are closed and kept in the dark. Luminescence, the color of which depends on the added dye, is observed immediately when 50 mL of the solutions A and B are mixed.

fluorescent dye	color	time of light emission	
		DNPO	TCPO
(I)	blue	ca. 12 min	> 3 h
(II)	green	ca. 15 min	> 3 h
(III)	red	ca. 2 min	ca. 30 min

If necessary, DNPO and TCPO can be prepared as follows: a solution of 0.1 g dinitrophenol (19.7 g) or trichlorophenol (18.4 g) in anhydrous toluene is stirred with 0.1 g freshly distilled triethylamine (10.1 g); the temperature must be kept below 10 °C. 55 mmol (7 g) oxalyl chloride is now added dropwise and the mixture left overnight at room temperature. After vacuum distillation of the reaction mixture a yellow residue is obtained which is stirred for 15 minutes with dichloromethane (in the case of DNPO) or light petroleum (TCPO). The precipitate, which is now white, is washed several times with a little cold solvent, dried for about an hour in a vacuum and recrystallised from the above solvents. The yields are about 38 % (DNPO) and 65 % (TCPO); the purity can be checked by measuring the melting points or the IR spectra.[1,2]

Explanation

The exact reaction mechanism is not known. It can however be assumed that the oxalate esters are first oxidised by the H_2O_2 to peroxyoxalate, which is then converted to the dioxethanedione. This forms a charge-transfer complex with the dye, and the complex decomposes to give CO_2 and the dye in an excited state. Light is emitted when the dye molecule returns to the ground state.

The catalytic action of the sodium salicylate can be clearly shown by the increase in brightness of the chemoluminescence when tiny amounts of the basic salt are added to the catalyst-free reaction mixture. The light intensity decreases considerably when the reaction mixture is cooled in an ice bath and increases when it is warmed in a water bath.

Waste Disposal

Any remaining organic solvents are transferred to the corresponding waste containers, while the reaction mixture itself is treated with soda solution, highly diluted and then poured down the drain.

References

1 A. G. Mohan, *J. Chem. Educ.,* **1974,** *51,* 528.
2 D. Potrawa, A. Schleip, *MNU,* **1983,** *36,* 284.

Hemoglobin Chemoluminescence

Luminiferous animal organisms are well known. Examples include the small sea animal *Noctiluca miliaris*, the firefly *Lampyris noctiluca* and the lamp jellyfish *Pelagia noctiluca*. The amazing light emission produced by these species is due to luciferase catalysts, whose coferment luciferin (molecular formula below) can regulate the bioluminescence processes by means of its action in redox and light transfer cycles.

These natural examples of chemoluminescence processes inspire us to create new laboratory experiments, another of which will be demonstrated below.

Safety Measures

Protective gloves and safety glasses must be worn!

Apparatus

Mortar and pestle, 1-L glass or polythene beaker, 1-L bottle with stopper, safety glasses, protective gloves.

Chemicals

Bovine hemoglobin (Sigma Chemical Co., St. Louis, MO) (or blood), luminol (5-amino-1,2,3,4-tetrahydrophthalazine-1,4-dione), sodium perborate tetrahydrate $NaBO_3 \cdot 4\ H_2O$, $Na_3PO_4 \cdot 12\ H_2O$, icing sugar, distilled water.

Experimental Procedure

0.2 g of luminol are mixed with 4.0 g of hemoglobin, 4.0 g $NaBO_3 \cdot 4\ H_2O$, 30 g $Na_3PO_4 \cdot 12\ H_2O$ and 30.0 g of icing sugar in a mortar and the mixture ground to give a fine powder. If fresh blood is to be used this must be added directly prior to the demonstration. 4 g of the mixture are now added to the liter bottle, which is half full of distilled water (add blood now if necessary) and shaken well.[1]

The room is darkened and the solution poured in a thin stream into the large beaker from a height of about 1 meter: the luminescence is observed as a brilliant blue color.

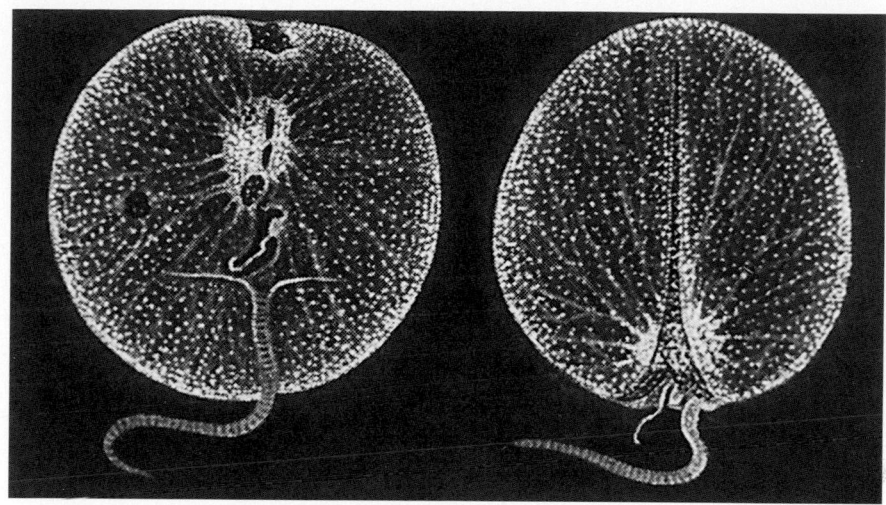

Noctiluca miliaris shown form both sides

Luciferin

Explanation

Like the structurally related chlorophyll, hemoglobin is an extremely good fluorophore, which gives up its activation energy in the form of blue light quanta on returning from the electronically excited state to the ground state. For further information see Experiments 65 et seq.

Waste Disposal

The mixture is diluted with a large amount of water and poured down the drain.

Reference

1 D. B. Phillips, *J. Chem. Educ.,* **1993**, *70*, 773.

74

Developing a Picture with Light

"It will be! the mass is working clearer!
Conviction gathers, truer, nearer!
The mystery which for Man in Nature lies
We dare to test, by knowledge led;
And that which she was wont to organize
We crystallize, instead.

Johann Wolfgang von Goethe, "Faust, The Second Part";
Laboratory

Apparatus

Flashlight or (preferably) halogen lamp, 600-mL beaker, glass rod, 10-mL and 5-mL measuring cylinders, two 150-mL beakers, cell $150 \times 10 \times 15$ mm made of glass or of plexiglass, stencil for covering cell (pattern cut from cardboard), protective gloves, safety glasses.

Chemicals

$Fe(NO_3)_3 \cdot 9\ H_2O$, oxalic acid, $K_3[Fe(CN)_6] \cdot H_2O$, Triton®-X-100 (nonionic surfactant), Cab-O-Sil® (highly dispersed fumed silicon dioxide).

Solution A

1.2 g $Fe(NO_3)_3 \cdot 9\ H_2O$ in 100 mL of water

Solution B

0.8 g oxalic acid in 100 mL of water

Solution C

10 mL 3 % $K_3[Fe(CN)_6] \cdot H_2O$ solution

Experimental Procedure

11 g of Cab-O-Sil are placed in a 600-mL beaker and the solutions A, B and C added. The mixture is stirred well until a uniform paste is formed, and the 3 mL Triton-X-100 stirred in.

   The thick yellow paste is placed in the cell, which is covered with the cardboard stencil from which the pattern has been cut and illuminated for about 5 seconds using the halogen lamp.[1] On removing the stencil the pattern is clearly seen in blue against the yellow background (colored figure 28). If the solutions are exposed to bright light after mixing the color of the mixture changes from the yellow of the iron oxalate complex to the dark blue color typical of Prussian Blue.

Explanation

On exposure to light Fe^{3+} is reduced to Fe^{2+}, which then reacts with $[Fe(CN)_6]^{3-}$ to give Prussian Blue.

$$2\,[Fe(C_2O_4)_3]^{3-} \xrightarrow{\;h\cdot v\;} 2\,Fe^{2+} + 2\,CO_2 + 5\,C_2O_4^{2-}$$

$$K^+ + Fe^{2+} + [Fe(CN)_6]^{3-} \rightarrow K[Fe(III)Fe(II)(CN)_6]$$

Waste Disposal

The residue is placed in the container used for heavy metal waste.

Reference

1 W. H. Batschelet, *J. Chem. Educ.,* **1986**, *63*, 435.

Where There Is Light, There Is Also Shadow; Experiments with UV Light

The known is finite, the unknown infinite; spiritually we find ourselves on a tiny island in the middle of a boundless ocean of the inexplicable.
   It is our task, from generation to generation, to drain a small amount of additional land.

Thomas Huxley

However, the path leading to the castle was lined on both sides ... with beautiful trees bearing all sorts of fruits, and always three trees on each side bearing lanterns, inside all the lamps already lit by a beautiful maiden ... in a blue gown with a wonderful torch. It was so magnificent and masterful to observe that, despite the pleas, I lingered rather long.

Johann Valentin Andreae

List number 5 – six undershirts, six undershorts, six handkerchiefs – has always left the investigators wondering, especially because of the complete absence of socks.

Woody Allen

Apparatus

UV lamp (254 nm), self-sealing plastic freezer bag, fluorescent TLC aluminum foil, glass rod, safety glasses, protective gloves.

Chemicals

Distilled water, 4-aminobenzoic acid.

Experimental Procedure

The light from the UV lamp is allowed to illuminate the fluorescent TLC aluminum foil, which takes on a yellow-green color. No change is observed when the freezer bag containing about 500 mL of water is held between the lamp and the TLC foil. About 1 g of the 4-aminobenzoic acid is added to the water and stirred until it dissolves. If the freezer bag is now again held between the lamp and the foil, the yellow-green color is no longer visible, the foil appearing to be dark blue.[1]

Explanation

4-Aminobenzoic acid absorbs light in the ultraviolet region of the spectrum.

The way in which light can destroy dyestuffs, plastics, fibres and the skin is well known, and the development of substances which can absorb UV light has been the subject of much research. These UV filters, for example for use in sun creams, should not themselves be colored but should be stable to light; they must be non-toxic, non-volatile and water-resistant. They should also be capable of filtering out the dangerous short wavelength UV light but not that part of the spectrum which tans the skin. Examples of such substances are glyceryl-4-aminobenzoates and 2-hydroxybenzophenones, substances which are related chemically to the 4-aminobenzoic acid used here.

Waste Disposal

The aqueous solution of 4-aminobenzoic acid can be poured down the drain.

Reference

1 E. Carberry, T. Gonella, R. Eliason, *J. Chem. Educ.,* **1989,** *66,* 1041.

A Blueprint

Since 1929 I have continuously studied the development and discoveries of the past hundred years in the individual disciplines. Even if the enormous specialization meant that I was not able to examine every corner, I nevertheless understood their significance as well as the best of them.

Salvador Dali (July 30, 1941)

Where nature ceases to permit new forms to come into being, there man begins to create, starting with the natural things and with the help of this very nature, an infinite variety of forms.

Leonardo da Vinci

Apparatus

2 developing dishes, glass or plastic plate, halogen lamp, various flat stencils, filter paper, safety glasses, protective gloves.

Chemicals

Distilled water, 1 mol/L hydrochloric acid, a solution of 10 g $K_3[Fe(CN)_6] \cdot H_2O$ and 13 g ammonium iron(III) citrate in 250 mL of water.

Experimental Procedure

A smooth white filter paper is impregnated with the solution of $K_3[Fe(CN)_6]$ and ferric ammonium citrate. The filter paper is laid on the glass plate and a stencil (e.g. cutout Micky Mouse figure, razor blade etc.) laid on the paper. The plate is now illuminated from above with a halogen lamp for about 10 seconds. The parts of the filter paper which are exposed to light turn blue (because of the formation of the well-known Prussian Blue) while the remainder stays a yellowish white. The negative so obtained is fixed by first immersing it in dilute hydrochloric acid for about 5 minutes and then washing it with running water from the tap to remove all traces of chemicals.[1]

Explanation

The citrate ion reacts similarly to the oxalate ion (see Experiment 74). Carbon dioxide is also formed here.

Waste Disposal

The residues should be transferred to the container used for heavy metal waste.

Reference

1 F. Bukatsch, O. P. Krätz, G. Probeck, R. J. Schwankner, *So interessant ist Chemie,* Aulis Verlag, Köln, **1987,** 67.

Photochemical Reduction
of a Thiazine Dye

We have hitherto seen the physiological colors displayed in the after-vision of colorless bright objects, and also in the after-vision of general colorless brightness; we shall now find analogous appearances if a given color be presented to the eye: in considering this, all that has been hitherto detailed must be present to our recollection.

The impression of colored objects remains in the eye like that of colorless ones, but in this case the energy of the retina, stimulated as it is to produce the opposite, color, will be more apparent.

Johann Wolfgang von Goethe, "Theory of Colors"(1810)

One must classify things not from without but from within.

Charles Ferdinand Ramuz (1878–1947)

Apparatus

600-mL beaker, 300 W lamp, black paper to wrap around the beaker, safety glasses, protective gloves.

Chemicals

Solution of 2 g $FeSO_4 \cdot 7 H_2O$ in 500 mL of water (or an equivalent amount of Mohr's salt), 20 % sulfuric acid, 0.02 g of the dyestuff Lauth's violet (thianine).

Experimental Procedure

480 mL of the iron sulfate solution, 10 mL of 20 % sulfuric acid and 10 mL of the dyestuff solution are mixed in the beaker; the mixture turns violet. The contents of the beaker are now illuminated with light from the lamp, which is placed about 4 cm above the surface. Decolorization is practically instantaneous. The lamp is removed and the beaker wrapped in black paper: the violet color is regenerated after a few seconds.

Structural formula of the 3,6-diaminophenthiazinium cation

Explanation

3,6-Diaminophenthiazine (see structural formula above), which was first pre-pared in 1876 by *Lauth* from 1,4-diaminobenzene and hydrogen sulfide, forms a violet cation in acidic solution.

Waste Disposal

The mixture is neutralized with milk of lime, diluted with water and poured down the drain.

Whiter Than White

Pure water crystallized to snow appears white, for the transparence of the separate parts makes no tansparent whole. Various crystallized salts, when deprived to a certain extent of moisture, appear as a white powder. The accidentally opaque state of a pure transparent substance might be called white; thus pounded glass appears as a white powder. The cessation of a combining power, and the exhibition of the atomic quality of the substance might at the same time be taken into the account.

Johann Wolfgang von Goethe, "Theory of Colors" (1810)

Apparatus

Quartz lamp, metal or plastic dishes which do not fluorescence, darkroom, protective gloves, safety glasses.

Materials

Samples of textiles and paper, various washing powders.

Experimental Procedure

The textile and paper samples are placed in the dishes with the various washing powders and checked for blue fluorescence in a darkroom using the quartz lamp. If the washing powder contains whiteners the surface of the samples which have come in contact with the powder shine abright white.

Explanation

The whitener which is added to the washing powder converts the UV light into visible light, so that the treated surface appears whiter. This additional light radiation, which is light blue in color, can be realised with the help of many organic molecules with delocalised π-systems (for an example see below).

Whitener molecule based on the diphenylurea structure

A Simple Luminophore

In Großenhain in Saxony there lived in 1676 a certain official, who was educated, curious, and talented, by the name of Balduin, *... a close acquaintance of* D. Früben, *at that time the doctor; these two managed to discover the Spiritum mundi, in particular how they could capture it with a convenient magnet, and how they could use it. For this purpose they took chalk, dissolved it in Spiritu nitri, abstracted it ad siccitatem, and exposed the residue to the air, which then absorbed water. This they extracted, and they designated the water as a Spiritum mundi, of which a portion of 12 Gr. must suffice, and both high and low made use of it.*

... In the course of this work it transpired on one occasion that they abstracted the Spiritum nitri too vigorously, with the result that something yellow was deposited in the neck of the retort; after they had ... broken the latter he threw the neck away ... in a dark place, and became aware that it glowed like an ember ... and that this light disappeared again into darkness, and acquired light anew from the sunlight. He brought this not only to privy council director Baron von Friesen, *but also to almost all the distinguished ministers in Dresden, and finally to me as well.*

This report is provided by *Johannes Kunckel,* groom of the chamber of the Elector of Saxony, in his *"Laboratorium Chymicum".*[1]

As early as 1602 the shoemaker *V. Casciaroli* from Bologna had produced a luminophore, which became famous as "Bologna phosphorus", by heating barium sulfate with flour to red heat.

Apparatus

Quartz lamp (for UV with a nickel oxide filter to cut out visible light), spectroscope, darkroom, spatula, mortar and pestle, porcelain dish, protective gloves, safety glasses.

Chemicals

Magnesium bromide crystals, tin(II) chloride,

Experimental Procedure

About 1 g $MgBr_2 \cdot 6 H_2O$ is ground to a powder in the dark; no luminescence is observed. A little $SnCl_2$ is now added and the mixture ground well. After a short time golden yellow fluorescence is observed when the powder is irradiated with UV light. With the help of a spectroscope this fluorescence can be shown to consist of a continuous emission in the long wavelength range (red to green).

If magnesium bromide is not available it can readily be prepared by shaking bromine water with magnesium powder until the color of the bromine has disappeared and the gas evolution has ceased (well-ventilated hood). The colorless filtrate is evaporated in a crystallising dish in a water bath until crystallisation commences, cooled, filtered and the crystals dried in an exsiccator over solid NaOH.

Explanation

Luminophores are substances which light up in certain colors, depending on their chemical structure, when irradiated with visible light, X-rays or electrons. Thus the incorporation of Sn^{2+} ions in the crystal lattice of $MgBr_2$ or its hydrate induces defect sites which cause the light emission responsible for the color observed. Luminophores can be "tailormade" by the selective introduction of such *"activators"* (generally alkaline earth cations) into the crystal lattice.

Waste Disposal

The substances are washed down the drain with water.

Reference

1 H. W. Prinzler, *Phosphorus, Sulfur, Silicon,* **1993,** *78,* 2.

80

The Setting Sun

May 6, P.M. – This is the hour for strange effects in light and shade – enough to make a colorist go delirious – long spokes of molten silver sent horizontally through the trees (now in their brightest tenderest green), each leaf and branch of endless foliage a lit-up miracle, then lying all prone on the youthful-ripe, interminable grass, and giving the blades not only aggregate but individual splendor, in ways unknown to any other hour. I have particular spots where I get these effects in their perfection. One broad splash lies on the water, with many a rippling twinkle, offset by the rapidly deepening black-green murky-transparent shadows behind, and at intervals all along the banks. These, with great shafts of horizontal fire thrown among the trees and along the grass as the sun lowers, give effects more and more peculiar, more and more superb, unearthly, rich and dazzling.

Walt Whitman, "Specimen Days" (1882)

Apparatus

Slide projector, screen, measuring cylinder, large flat-bottomed flask (1-L or 2-L), beakers, glass rods, dropping pipettes, room which can be darkened, safety glasses, protective gloves.

Chemicals

5 % Hydrochloric acid (water and concentrated hydrochloric acid in a ratio of about 7:1), 2 % sodium thiosulfate solution (with respect to the anhydrous salt).

Experimental Procedure

A large flat-bottomed flask is filled with the 2 % sodium thiosulfate solution and stood in front of the projector lens so that the light from the projector passes through the center of the flask; the screen is placed about 2 metres from the projector. The distance from screen to projector and the focussing of the lens is now adjusted to produce a white circle on the screen; this represents the sun.

The dilute hydrochloric acid is now added slowly from a dropping pipette with continuous careful stirring. A slight light scattering effect is observed from the side; this increases steadily, and the *Tyndall* cone produced is first a milky yellow, then yellowish. The circle on the screen changes from bright white via diffuse yellow to orange-red, and then to a dark twilight hue.

Explanation

The sulfur allotrope S_6 is mainly formed in the reaction between sodium thiosulfate hydrochloric acid.

This elemental sulfur is initially produced in colloidal form, and the light scattering effect first studied and rationalised in detail by the English physicist *J. Tyndall* (1820 – 1893) can be observed. Shorter wavelength light is scattered more strongly than longer wavelength light. The picture of the "setting sun" on the screen steadily becomes more diffuse and redder, until it disappears in darkness. The illusion of the setting sun is improved if the projector is slowly lowered.[1, 2]

Waste Disposal

The solution can be poured down the drain after neutralisation.

References

1 B. Z. Shakhashiri, *Chemical Demonstrations, A Handbook for Teachers of Chemistry,* University of Wisconsin Press, Madison, London, **1983, 3,** 353.
2 F. Bukatsch, O. P. Krätz, G. Probeck, R. J. Schwankner, *So interessant ist Chemie,* Aulis-Verlag, Köln, **1987,** 263.

81

Mercury Beating Heart

Nec tamen undique corporea stipata tenentur
omnia natura; namque est in rebus inane.
Quod tibi cognosse in multis erit utile rebus
nec sinet errantem dubitare et quaerere semper
de summa rerum et nostris diffidere dictis.
Qua propter locus est intactus inane vacansque.
Quod si non esset, nulla ratione moveri
res possent; namque officium quod corporis exstat,
officere atque obstare, id in omni tempore adesset
omnibus, haud igitur quicquam procedere posset,
prinicpium quoniam cedendi nulla daret res.

Titus Lucretius Carus, "De Rerum Natura"

These ideas of the Greek philosophers *Demokrit* and *Epikus* are emphatically
translated in *"The Way Things Are"*:

But not all bodily matter is tight-packed
By natures law, for there's a void in things.
This knowledge will be useful to you often,
Will keep you from the path of doubt, from asking
Too many questions on the sum of things,
From losing confidence in what I tell you.
By void I mean vacant and empty space,
Something you cannot touch. Were this not so,
Things could not move. The property of matter,
Its most outstanding trait, is to stand firm
Its office to oppose; and everything
Would always be immovable, since matter
Never gives way.

Safety Measures

Chromates and dichromates are highly toxic and carcinogenic. Protective gloves should be worn. Mercury is also highly poisonous.

Apparatus

Overhead projector, watchglass 8 to 10 cm in diameter, iron nail or pin, safety glasses, protective gloves.

Chemicals

Mercury, concentrated H_2SO_4, 30 % H_2SO_4, 0.1 mol/L $KMnO_4$ solution or 0.1 mol/L $K_2Cr_2O_7$ solution.

Experimental Procedure

Enough mercury to form a drop about 2 cm in diameter is placed on the watchglass, which in turn is placed on the overhead projector. The mercury drop is covered with 30 % sulfuric acid and a few drops of the solution of the oxidising agent are added (two or three crystals of the solid salt can also be used). The iron nail is now placed at the edge of the watchglass so that its point just touches the mercury. Up to 2 mL of the concentrated sulfuric acid are now added dropwise. At once the mercury begins to switch and oscillates to and fro. Sometimes the oscillating figure resembles a heart.

Explanation

Electrons flow from the base metal iron to mercury, so that the surface tension of the latter changes. The resulting rhythmic movements cause the mercury to lose contact with the iron, so that the electrons flow into the sulfuric acid and the oxygen-containing salts; the surface tension of the metallic mercury then changes once more, so that it again comes into contact with iron and the cyclic process is repeated.[1]

Waste Disposal

The reaction solutions can be re-used; if no longer required, the salt solutions are transferred to the container for highly toxic waste. The sulfuric acid is diluted with water. If mercury is to be disposed of, it should be absorbed using Chemizorb® (Merck) and placed in a plastic container labelled POISON.

Reference

1 D. Avnir, *J. Chem. Educ.* **1989,** *66,* 211.

Gallium Beating Heart

The European-American-Israeli gallium experiment GALLEX has probably provided the first direct information on the basic mechanism for energy generation in the interior of the sun. The most important energy-generating process is the fusion of hydrogen to give helium, which is accompanied by the formation of neutrinos. GALLEX is based on the conversion of gallium to germanium:

$$^{71}_{31}\text{Ga} + \nu_e \rightarrow {}^{71}_{32}\text{Ge} + e^-$$

Neutrinos can be detected at energies above 233 keV. After the completion of the first phase of GALLEX on April 29th 1992 it appears that neutrinos have in fact been detected, thus providing an initial prerequisite for possible corrections to the previous sun model.[1]

Chemically it is of interest that both elements involved in the GALLEX experiment, gallium and germanium, provided the first evidence that the Periodic Table of the Elements as proposed by *D. I. Mendeleev* was correct. It was presented publicly on March 1st 1869 in St. Petersburg under the title *"The Arrangement of a System of the Elements based on their Atomic Weight and Chemical Structure"*. The element gallium, the existence and many of whose properties had been predicted by *Mendeleev*, was discovered in 1875 by *P. E. Lecoq de Boisbaudran*, while the element germanium, which he had also described in detail, was discovered by *C. A. Winkler* a few years later (1886).

Safety Measures

Potassium dichromate is highly toxic and carcinogenic. Protective gloves must be worn when this experiment is being carried out.

Apparatus

Petri dish, iron nail or wire, protective gloves, safety glasses.

Chemicals

Gallium, 30 % sulfuric acid, concentrated sulfuric acid, 3 % aqueous $K_2Cr_2O_7$ solution.

Experimental Procedure

A small amount of gallium (m.p. 29.8 °C) is heated to about 40 °C and poured into the Petri dish. The liquid metal is now covered with a mixture of the 30 % sulfuric acid and the potassium dichromate solution and the iron nail (or wire) is laid on the edge of the Petri dish so that its point just touches the gallium. Concentrated sulfuric acid is now added dropwise; the gallium metal begins to

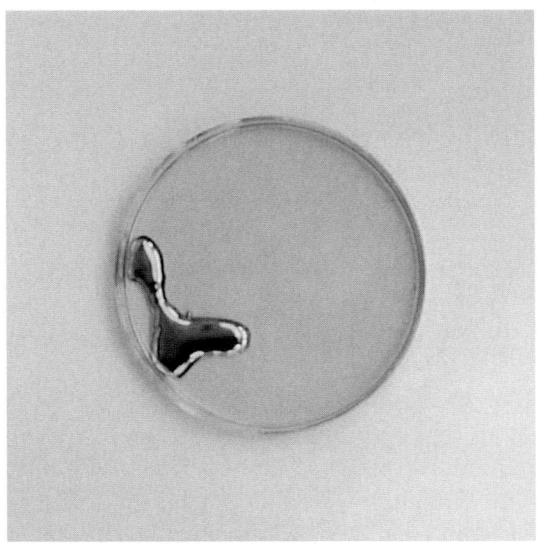

Gallium beating heart

move rhythmically and takes up bizarre forms, which at times can resemble a heart (see figure). The process can be kept going for up to 10 minutes by addition of a further few drops of sulfuric acid or of a little potassium dichromate solution. The metal remains liquid for a considerable time.[2]

Explanation

The process resembles that described in experiment 81; though gallium is not a noble metal, it is oxidised only very slowly, so that this experiment can be used as a less toxic alternative to the mercury heart.

Waste Disposal

The contents of the Petri dish are transferred to the container used for highly toxic waste.

References

1 R. Vaas, *Naturwissenschaftliche Rundschau*, **1993**, *46*, 56.
2 J. L. Ealy Jr., *J. Chem. Educ.*, **1993**, *70*, 491.

How to Make Batteries from Fruit and Vegetables

Experiment A: A Lemon Battery

Apparatus

Flashbulbs, 6 copper and 6 zinc electrodes (10 × 1 cm), capacitor (16 V, 470 µF), 4 long cables, 5 short cables, 3 alligator clips, high-impedance voltmeter, safety glasses, protective gloves.

Materials

Six lemons.

Experimental Procedure

Before starting the experiment the audience should be shown that the capacitor is uncharged. Copper and zinc electrodes are stuck into the six lemons and connected in series (cable from the Zn to the Cu electrode) and with a voltmeter; the Cu electrode of the first lemon must be attached to the positive pole, the Zn electrode of the last lemon to the negative pole. It can be seen clearly that a voltage of about 5 V is generated. The voltmeter is now replaced by the capacitor, the Cu electrode of the first lemon being attached to the positive side, the Zn electrode of the last to the negative side. The capacitor is charged for about a minute. A flash is observed if a flashlamp is brought into contact with the capacitor with the help of a piece of wire.

Experiment B: A Potato Battery

Apparatus

Battery-driven clock with second hand, 5 copper and 5 zinc electrodes (10 × 1 cm), capacitor (40 V, 2.2 mF), 4 short and 4 long cables, alligator clips, high-impedance voltmeter.

Materials

Five potatoes.

Experimental Procedure

Copper and zinc electrodes are stuck into two potatoes and connected according to experiment A. Again voltage is generated, which can be increased with further potatoes. After the capacitor is charged, it can be used to switch on a small electric watch. With two potatoes the watch runs for approximately 40 seconds, with three potatoes for approximately 60 seconds, five potatoes supply enough energy for over 80 seconds (if the capacitor has been charged for one minute according to experiment A).

Both batteries also run with apples, oranges, and other fruits.

Explanation

The weakly acid fruit and vegetable juices and the zinc electrode form a galvanic element, the cell reaction of which can be described as follows (eqns. 1 and 2):

(1)   (−) pole: $Zn \rightarrow Zn_{aq}^{2+} + 2e^-$

(2)   (+) pole: $2H_{aq}^+ + 2e^- \rightarrow H_2$

The resulting voltage is directly proportional to the maximum work, as is demonstrated in different ways in the two experiments. The copper electrodes can be replaced by platinum or other noble metals, as they serve only to collect electrons.

Additional information can be found in the recently-published experiment *"A Lemon-Powered Clock"*.[1]

Reference

1 T. M. Letcher, A. W. Sonemann, *J. Chem. Educ.*, **1992**, *69*, 157.

Colors around the Cathode

A picture is not something that is completely planned and fixed from the outset. As one works on it, it changes in the same degree as ones thoughts. And when it is complete, it continues to change, corresponding to the frame of mind of he who is at that moment viewing it. A picture lives out its own life as a living creation, and it undergoes the same changes that we are subject to in everyday life. This is quite natural, because a picture has life only through the person who observes it.

Pablo Picasso

Apparatus

Two Petri dishes, insulated copper wire, banana plugs, alligator clips, glass plate, polished nickel-plated sheet iron (about 10 × 15 cm) as support, source of DC current (10V), hair dryer, spatula, measuring cylinder, dropping pipettes, safety glasses, protective gloves.

Chemicals

5 % sodium chloride solution, ethanol, 1 % alcoholic solutions of bromothymol blue and phenolphthalein, filter paper.

Experimental Procedure

(I)   The white filter paper is impregnated with the NaCl solution to which have been added a few drops of the indicator bromothymol blue and a few milliliters of ethanol, dried almost completely with the hair dryer and placed on the glass plate. The DC current source (10V) is brought into contact with the paper by means of alligator clips, copper wire and banana plugs in such a way that the poles are about 1 cm apart. The paper, which is colored yellow by the indicator, turns deep blue in the neighbourhood of the cathode.

(II)   If the bromothymol blue is replaced by an alcoholic phenolphthalein solution, the paper is initially colorless and turns red in the vicinity of the cathode.

   If one intends to write or draw electrolytically, the polished nickel plated sheet iron is placed underneath the filter paper and connected to the positive pole of the DC source by means of an alligator clip made of metal. The banana plug is connected to the negative pole and can be used to produce a blue or red drawing on the filter paper.[1]

Explanation

During the electrolysis of an aqueous sodium chloride solution elemental chlorine is formed at the anode from the chloride ions, while at the cathode H_3O^+ ions are reduced to molecular hydrogen. The area around the cathode contains an aqueous solution of the Na^+ and OH^- ions, which determine its basicity and cause the indicator to change color. It is thus possible to generate different color effects in the area around the cathode by changing the indicator used.

Waste Disposal

Solutions which are no longer required can be poured down the drain.

Reference

1 F. Bukatsch, O. P. Krätz, G. Probeck, R. J. Schwankner, *So interessant ist Chemie*, Aulis-Verlag, Köln, **1987**, 60.

85

How to Turn Aluminum into Hoarfrost

Nature! We are encircled and enclasped by her – powerless to depart from her, and powerless to find our way more deeply into her being. Without invitation and without warning she involves us in the orbit of her dance, and drives us onward until we are exhausted and fall from her arm.

Eternally she creates new forms. What now is, never was in time past; what has been, comes not again – all is new, and yet always it is the old.

We live in the midst of her, and yet to her we are alien. She parleys incessantly with us, and to us she does not disclose her secret. We influence her perpetually, and yet we have no power over her.

It is as if she founded all things upon individuality, and she recks nothing of individuals. She builds for ever, and destroys for ever, and her atelier is inaccessible.

Johann Wolfgang von Goethe, "Nature: Aphoristic"

Safety Precautions

Mercury and its salts are highly toxic.

Apparatus

Bottle with dropping pipette, small beaker, safety glasses, protective gloves.

Chemicals

Sheet aluminum, an aqueous solution of mercury(II) chloride which is saturated at room temperature (ca. 7 g $HgCl_2$ in 100 mL water), 1 mol/L HCl.

Experimental Procedure

A few milliliters of the dilute hydrochloric acid are added to the mercury chloride solution; a few drops of the resulting mixture are carefully placed on the sheet aluminum, which has previously been polished until it is shiny. At first the surface of the aluminum becomes dull and mat. Soon a white layer resembling a mold fungus is formed on the surface, and after a while needles begin to grow, producing a greyish-white "lawn".

Explanation

A local galvanic element is formed at the surface of the sheet aluminum, whose protective oxide layer is destroyed by the acid, and the Hg^{2+} ions are reduced by the aluminum to metallic mercury. In turn the aluminum is converted in the presence of the acid to $[Al(H_2O)_6]Cl_3$, the white needles of which form the "hoarfrost". The grey color is due to the presence of metallic mercury.

Waste Disposal

Mercury and mercuric chloride should be absorbed using Chemizorb® (Merck) and transferred to a waste container labelled POISON.

86

How to Clean Silver Cutlery

And who shall stand when he appeareth?
For he is like a refiner's fire,
And like fuller's soap;
And he shall sit as a refiner and purifier of silver;
And he shall purify the sons of Levi,
And purge them as gold and silver.

The Book of Malachi 3, 2–3

Apparatus

Large flat glass dish suitable for accommodating the silver cutlery, 800-mL beaker, safety glasses, protective gloves.

Reacting Substances

Tarnished silver cutlery, aluminum foil, 15 % soda solution (with respect to anhydrous soda), NaCl.

Experimental Procedure

The glass dish is lined with aluminum foil, on which the tarnished silver cutlery is placed. The soda solution is poured in and left to stand for a few minutes (not more than 10). The cutlery is now washed with distilled water: it is completely clean and shiny. The same effect is obtained when table salt is poured on to the tarnished cutlery and the latter wrapped in aluminum foil and rubbed hard.

Explanation

In alkaline solution aluminum is a reducing agent with respect to water and the Ag^+ ions present on the surface of the tarnished silverware; it is itself oxidised to Al^{3+} ions, which first form $Al(OH)_3$ with the hydroxide ions of the soda solution. The uptake of a further OH^- ion gives the water-soluble $[Al(OH)_4]^-$ complex. The hydrogen set free reinforces the reduction process, so that the silver cutlery gets its former shine back.

Waste Disposal

The contents of the cleaning bath are poured down the drain and the aluminum foil disposed of with the household waste.

87

Experiments with Liquid Nitrogen

Everything has its science;
with the exception of catching fleas:
That is an art.

Dutch proverb

Oxygen was discovered three times: first by *Carl Scheele* as *"fire air"*, 1774 by *Joseph Priestley* as *"dephlogisticated air"* and in the same year by *Antoine Laurent Lavoisier* as *"the air of life"*.

During my sojourn in England from September 1787 to February 1788 I had the opportunity to become acquainted in Birmingham with Priestley, Withering, *and* Keir. Priestley *lived ... in utter seclusion two English miles from Birmingham in a country estate. He had a very well-equipped laboratory and in particular a very good set of apparatus for studying airs. It was a joy to my heart to be able here to see various pieces of his apparatus for myself, things which I already knew from his writings. He also had the kindness to show me some of the glass tubes made of flint glass with blackened inner surfaces, tubes in which he had carried out the experiment with inflammable air in order in this way to demonstrate the reduction of lead, one of the constituents of the glass, as a way of proving that phlogiston and inflammable air are one and the same thing.*

J. F. A. Göttling,
"Almanach oder Taschenbuch für Scheidekünstler und Apotheker" (1789)

On the occasion of a visit to Paris during the year 1774, Priestley *reported (at* Lavoisier's *chalkboard) on his new form of air obtained, e.g., by heating mercury oxide in which a candle burns much more vigorously, i.e., about his "dephlogisticated air", and in April 1775* Lavoisier *was already able to present his report "concerning the nature of the principle that binds with metals during their calcination and increases their weight". This was* Priestley's *"dephlogisticated air" known from then on as "oxygène" or* "principe acidifiant" *which combines with metals and reveals itself as a material increase through the increased weight of metal calcinates, "oxides".*

Paul Walden, Geschichte der Chemie (1950)

The liquefaction of gases was first achieved by *James Prescott Joule* (1818–1889) and *William Thomson* (1824–1907). In 1892 *William Thomson* was raised to the peerage as *Lord Kelvin of Largs*. *Joule* and *Thomson* showed that the temperature of a gas which is allowed to expand freely falls somewhat. This phenomenon, the Joule-Thomson effect, is due to the very small intermolecular attraction between gas molecules. During the expansion a small amount of energy is taken up by the molecules which are diffusing apart in order to overcome the attractive force between them. Gases can be liquefied by making use of the Joule-Thomson effect. The boiling point of liquid nitrogen is −196 °C. For comparison, the record low on the earth's surface is −88.3 °C, measured on August 24th 1960 in Vostok in the Antarctic, while the lowest temperature ever measured in the atmosphere is −143 °C at a height of about 90 km.

Cooling in Liquid Nitrogen

Safety Precautions

Safety glasses must be worn! Avoid skin contact with liquid nitrogen, as it can cause severe skin burns.

Materials

Dewar vessels, hammer, wire, crucible tongs, rubber tubing, rubber ball, balloon, apple, pear, flower, safety glasses, leather gloves.

Chemicals

Liquid nitrogen.

Experimental Procedure

A piece of rubber tubing is slowly dipped into liquid nitrogen. It loses its elasticity on cooling and shatters into many pieces on being hit with a hammer when cold. An apple or a rubber ball behave in a similar manner; these can be dipped into the liquid nitrogen with the help of a wire loop.

Now the flower is cooled in liquid nitrogen. When it is allowed to fall on the table it shatters into a thousand pieces.

The balloon is blown up with air and then cooled in a Dewar vessel 10–20 cm in diameter. It first shrinks and then collapses when dipped further into the liquid nitrogen (using the crucible tongs). The gaseous air inside the balloon liquefies on cooling, so that there is an extreme decrease in volume. It is impressive to see how the balloon returns to its original shape and volume on warming up to room temperature.

Changes in Properties at Low Temperatures

Safety Precautions

For information on the dangers of dealing with liquid nitrogen see the previous section. HgI_2 is highly toxic: avoid skin contact.

Apparatus

Dewar vessels, test tubes, lead bell, safety glasses, protective gloves.

Chemicals

Liquid nitogren, sealed glass ampoules containing a) nitrogen dioxide and b) a 1:1 mixture of nitrogen dioxide and nitrogen monoxide; in addition HgI_2, sulfur, Pb_3O_4 (red lead).

Experimental Procedure

Prior to the experiment, HgI_2, sulfur and Pb_3O_4 are placed in test tubes, which are sealed. In each case the color becomes noticeably lighter when the test tubes are immersed in liquid nitrogen.

In the case of NO_2 the brown color disappears, and its dimer N_2O_4 condenses out as a colorless solid. The mixture of nitrogen dioxide and monoxide gives a blue liquid on cooling; this solidifies to a blue solid (N_2O_3) on further cooling.

A thin-walled lead bell is cooled for about 10 minutes in liquid nitrogen. When the bell is cold it is found to produce a much clearer brighter sound.

Waste Disposal

The ampoules can be used for further experiments.

88

Cigars Burn Better in *Liquid* Air!

He who smokes cigars and keeps pigeons
can watch his money flying through the air.

East Prussian proverb

Safety Precautions

Safety glasses must be worn! Do not let liquid air come into contact with the skin, as it can cause serious "burns". Liquid air can react explosively with organic compounds.

Apparatus

Dewar vessel, large porcelain dish (diameter 15 cm), cigarette lighter, safety glasses, protective gloves.

Materials

Liquid air, cigar.

Experimental Procedure

The porcelain dish is filled with liquid air and left to stand for about 10 minutes. Most of the nitrogen evaporates, and bluish oxygen-enriched liquid air remains. The cigar is dipped into this vertically for about 20 seconds, to that only one end absorbs liquid air. This cold end is now ignited with the cigarette lighter. The cigar burns very quickly with a bright flame.

Demonstration of the Meissner-Ochsenfeld Effect: A Hovering Superconductor

One does not predict the future, one creates it.

A. Sinov'ev

Just as we do not inquire for what useful purpose the birds sing, because they were created to sing and song is a delight for them, so we should also not ask why the human spirit strives to ascertain the secrets of heaven. Natural phenomena are so diverse, and heaven is so rich in hidden treasures, in order that the human spirit will never want for fresh nourishment.

Johannes Kepler, "Mysterium Cosmographicum"

In 1987 *J. G. Bednorz* and *K. A. Müller* were awarded the Nobel prize for physics for their discovery of the high temperature superconductivity exhibited by compounds with a perovskite structure.

Safety Precautions

Safety glasses must be worn. Avoid skin contact with liquid nitrogen, as it can cause severe burns.

Apparatus

Arc lamp or overhead projector, plastic tweezers, Dewar vessel, piece of plastic or polystyrene beaker, protective gloves, safety glasses.

Chemicals

Superconductor pellet made from Y_2O_3-$BaCO_3$-CuO, special magnet (Co-Sm), liquid nitrogen.

Preparation of the Superconductor

1.13 g Y_2O_3, 3.95 g $BaCO_3$ and 2.39 g CuO are mixed well and ground together carefully in a mortar. The mixture is heated for 12 hours at 950 °C in a corundum boat. Temperature control is extremely important and should be checked with a calibrated thermocouple. The mixture is allowed to cool to room temperature and again ground in the mortar; the fine powder is pressed into a pellet. The pellet is again heated to 950 °C for 12 hours in an oxygen atmosphere and allowed to cool slowly (100 °C per hour) to room temperature; it must remain in an oxygen atmosphere during the cooling process.[1]

Experimental Procedure

A plastic beaker (or similar article) is placed on the overhead projector in such a manner that the bottom of the beaker is visible. The pellet is cooled in liquid nitrogen and (with continued cooling) placed on the bottom of the beaker. The magnet is placed on top of it using the plastic tweezers. If the cooling is sufficient the magnet will float above the pellet. If the temperature of the superconductor exceeds the transition temperature, the magnet sinks on to the pellet. If the experiment is carried out in a larger room, a small paper flag can be

The magnet hovers in the air and rotates in the magnetic field

attached to the magnet which is then set spinning while it hovers. The floating, rotating magnet is then more easily visible for the audience (see figure).

Reference

1 K. Roth, *Chemie in unserer Zeit*, **1988**, 22, 30.

90

A Highly Exothermic Reaction

The German chemist *Victor Meyer* (1848–1897), the successor to *Robert Wilhelm Bunsen* in Heidelberg, developed a method for the determination of molecular masses via vapor density measurements. A sample of diphenylmethane always lay on his desk in summer. A colleague asked him why: his answer was *"Diphenylmethane melts at 26 °C. So if the sample melts I go swimming with my students".*

Safety Precautions

Skin contact with calcium oxide can cause burns!

Apparatus

250-mL beaker, thermometer (covering the range from −10 °C to +150 °C), protective gloves, safety glasses, spoon.

Chemicals

Pieces of quicklime (CaO), distilled water, dilute hydrochloric acid.

Experimental Procedure

About 100 g of CaO are placed in the beaker using a spoon. The thermometer is placed between the pieces so that its bulb touches the base of the beaker. 25 mL of distilled water are then poured on to the quicklime. The temperature rises immediately, reaching 95 °C in a few seconds and increasing to almost 115 °C. Steam is evolved during the slaking process and the volume of the solid increases considerably. The calcium hydroxide (slaked lime, $Ca(OH)_2$) formed has a pH of 12.4.

Explanation

The reaction of CaO with water, the slaking of quicklime, produces slaked lime, $Ca(OH)_2$, the ions of which are hydrated and take up a thermodynamically stable structure. This process involves a reaction enthalpy $\Delta H^0 = -81.9$ kJ/mol, and is accompanied by an increase in volume of a factor of three. The spontaneous heating of aqueous solutions is made use of in many ways; in most cases the solids which are slaked are not so aggressive as quicklime. Thus *packs used in first aid* for producing warmth contain anhydrous $CaCl_2$; this is present in the outer pack itself and the water (which contains a red dye) is in an inner bag. When warmth is required the package is pressed hard so that the inner bag bursts and its contents mix with the calcium chloride. The exothermic reaction, which is due to the hydration of the Ca^{2+} ions and (to a lesser extent) of the Cl^- ions causes the temperature to rise rapidly from 20 °C to over 70 °C; in some cases it can reach over 100 °C. The warmpacks normally contain 218 g $CaCl_2$ and 170 g H_2O.[1]

Waste Disposal

The solution should be neutralised, diluted with a large amount of water, and poured down the drain.

Reference

1 B. Z. Shakhashiri, *Chemical Demonstrations, A Handbook for Teachers of Chemistry,* University of Wisconsin Press, Madison, London, **1983**, *1*, 19.

91

Highly Endothermic Reactions

So cold, so icy, that one burns one's finger at the touch of him! Every hand that lays hold of him shrinks back! – And for that very reason many think him red-hot.

Friedrich Nietzsche (1844–1900), "Beyond Good and Evil"

Apparatus

200-ml beaker, thermometer (covering the range from $-50\,°C$ to $+50\,°C$), wooden board $10 \times 10 \times 2$ cm, glass rod, protective gloves, safety glasses.

Chemicals

$Ba(OH)_2 \cdot 8\,H_2O$, NH_4SCN, NH_4NO_3.

Experimental Procedure

15 g of barium hydroxide are mixed well with 5 g of ammonium thiocyanate in the beaker and the mixture placed on a wooden board, the surface of which is damp. Within a few seconds the smell of ammonia is clearly noticeable. The mixture of solids becomes liquid and its temperature falls within one to two minutes from about $+20\,°C$ to $-25\,°C$ or even below, as the thermometer shows; even after 10 minutes the temperature is still about $-20\,°C$. If the beaker is picked up, it is found to be frozen solid to the wood.[1]

Explanation

The solid reaction takes place according to equation (1):

(1) $Ba(OH)_2 \cdot 8\,H_2O(s) + 2\,NH_4SCN(s) \rightarrow Ba(SCN)_2 + 2\,NH_3(g) + 10\,H_2O$

This highly endothermic reaction can only occur because it is clearly entropy-determined. The disorder of the system increases because of the formation of a large number of single molecules, to that the entropy increases dramatically. The free enthalpy ΔG is thus negative, since the value of the product $T\Delta S$ is larger than the relatively high positive enthalpy; this leads to the decrease in temperature (eqn. (2)):

$$(2)\ \Delta G = \Delta H - T \cdot \Delta S$$

This cooling effect is made use of in cooling packs for first aid. The inner bag in the pack contains water which is colored blue, while the pack itself contains solid ammonium nitrate. When the cooling effect is required the inner bag is pressed hard so that it bursts and the water can mix with the salt. Commercial packs contain about 220 g NH_4NO_3 and an equal amount of water; the temperature falls from $+20\,°C$ to about $-7\,°C$, a temperature which is sufficiently low for cooling body organs.

Waste Disposal

Barium salts are toxic and must be placed in the container used for collecting toxic inorganic salts.

Reference

1 B. Z. Shakhashiri, *Chemical Demonstrations, A Handbook for Teachers of Chemistry*, University of Wisconsin Press, Madison, London, **1983**, *1*, 10.

92

An Eruption Caused by Zinc and Sulfur

The sun was risen upon the earth when Lot came unto Zoar. Then the Lord
rained upon Sodom and upon Gomorrah brimstone and fire from the Lord
*out of heaven; and he overthrew those cities; and all the plain, and all the
inhabitants of the cities, and that which grew upon the ground. But his wife
looked back from behind him, and she became a pillar of salt. And Abra-
ham gat up early in the morning to the place where he had stood before the*
Lord: *and he looked toward Sodom and Gomorrah, and toward all the
land of the plain, and beheld, and, lo, the smoke of the land went up as the
smoke of a furnace.*

The Book of Genesis 19, 23–28

Apparatus

Porcelain spoon, 100-mL beaker, gas burner, knitting needle, metal support
about 50 × 50 cm, large watchglass, safety glasses, protective gloves.

Chemicals

Zinc and sulfur powder, filter paper.

Experimental Procedure

20 g of zinc powder and 40 g of flowers of sulfur are mixed on a piece of filter
paper and the mixture placed in the beaker, which is covered with the watchglass. A sample is removed using the porcelain spoon and placed in the center
of the metal plate, which is in turn placed on a fire-resistant support. The mixture is set alight by touching it with the knitting needle, which has previously
been heated to red heat. The reaction commences after a short time. At first the
sulfur starts to melt, and then the "eruption" begins; it is accompanied by flames and hissing noises and white smoke is formed. A yellow solid is formed
which becomes white on cooling. This experiment can also be carried out using
three portions of the mixture placed next to one another on the metal support.

Explanation

Zinc reacts with sulfur in a strongly exothermic reaction to form zinc sulfide and with oxygen to give zinc oxide; the latter is yellow when hot but becomes white on cooling (*thermochromism*).

Waste Disposal

The reaction products are transferred to the container used for collecting less toxic inorganic waste.

93

Thermochromism

My nocturnal illumination is magnificent, indeed I prefer it to daylight ...
You must come sometime to see how the light sets off every object, how deep
black shadows frame all the pictures and are thrown upon the beams. You
can find that in almost all my still lifes, most of which I have painted at
night ...

G. H. Brassai, *"Conversations avec Picasso"* (1964)

Safety Precautions

Mercury salts are poisonous. Protective gloves must be worn.

Apparatus

200-mL beaker, two 100-mL beakers, glass rods, funnel with filter paper, spatula, protective gloves, safety glasses.

Chemicals

KI, $Hg(NO_3)_2$, or $HgCl_2$, CuCl, $AgNO_3$, distilled water, four round filter papers, filter paper for drying.

Experimental Procedure

5 g $Hg(NO_3)_2$ (or the same amount of $HgCl_2$) are stirred with 150 mL of water in the 200-mL beaker; the clear solution is separated from any precipitate which appears and treated with solid potassium iodide until the deep red precipitate initially formed dissolves to give a colorless solution. Half of this solution is placed in each of the 100-mL beakers. A 5 % $AgNO_3$ solution is added to the first half, a freshly prepared saturated CuCl solution to the second. The mixtures are allowed to stand for a while, decanted and filtered through filter paper. The precipitate from the first beaker is deep yellow, that from the second bright red. Both precipitates are dried by placing them between sheets of filter paper.

The yellow silver salt is spread on two round filter papers and the red copper salt on two others. The filter papers are fixed in position in an appropriate manner; one of each is heated with a hair dryer, while the other is left untreated as a control. The silver salt becomes deep red on heating and on cooling returns to its initial yellow color. The copper salt turns chocolate-brown or even black on heating and also returns to its initial color on cooling. These color changes are thus reversible (colored figure 29).

Explanation

Hg^{2+} salts react with I^- ions to give a dark red precipitate of HgI_2, which is converted on further addition of iodide ion to the colorless and highly water-soluble complex ion $[HgI_4]^{2-}$. This complex ion reacts with soluble silver salts to give the deep yellow $Ag_2[HgI_4]$, while Cu^+ ions produce a red precipitate of $Cu_2[HgI_4]$. Both substances undergo a complete color change on heating; in the examples used here the initial color returns on cooling. This type of thermochromism is reversible; the complex $Cu_2[HgI_4]$ turns brown-black at 70 °C, and is used industrially as a color indicator to check whether machine parts are becoming too hot when running. The thermochromism of zinc oxide from white to yellow is also reversible (see Experiment 92), while the color change of NH_4VO_3 from white to brown at 150 °C is irreversible, just as is that of $PbCO_3$ at 290 °C from white to yellow. Many other similar examples are known.

Waste Disposal

The residues from the $K_2[HgI_4]$ solution must be completely precipitated out (see above); the precipitates should be transferred to the container used for collecting toxic inorganic salts.

A Simple Experiment to Demonstrate the "Greenhouse Effect"

When I came to people, then I found them resting on an old infatuation: all of them thought they had long known what was good and bad for people.

And old wearisome business seemed to them all discourse about virtue; and he who wished to sleep well spoke of "good" and "bad" before retiring to rest.

This somnolence did I disturb when I taught that no one yet knows what is good and bad: – unless it be the creating one!

– It is he, however, who creates man's goal, and gives to the earth its meaning and its future: he only effects it that nothing is good or bad.

Friedrich Nietzsche, "Thus spoke Zarathustra"

The amount of carbon dioxide generated annually by burning fossil fuels and calcining limestone for the production of cement is increasing steadily (see figure). In spite of all the efforts made by discerning scientists, engineers and businessmen it is highly unlikely that this emission will cease in the near future. However, model calculations predict that if the concentration of CO_2 in the atmosphere rises to double its present value there will be an increase of between 1.5 and 4.5 °C in the temperature of the atmosphere at the earth's surface; this is likely to have a dramatic influence on vegetation, climate and the level of the oceans.

Apparatus

Two 250-mL beakers, two brass discs 0.5 mm thick and 63 mm in diameter, lamp with shade and 100 W bulb, coaxial Ni/CrNi thermocouple connected to both beakers, thermometer, glass plate, gas inlet tube, safety glasses, protective gloves.

Chemicals

CO_2 from a pressure cylinder, 250 mL of a 10 % NaOH solution which contains about 2.5 g $K_2S_2O_8$.

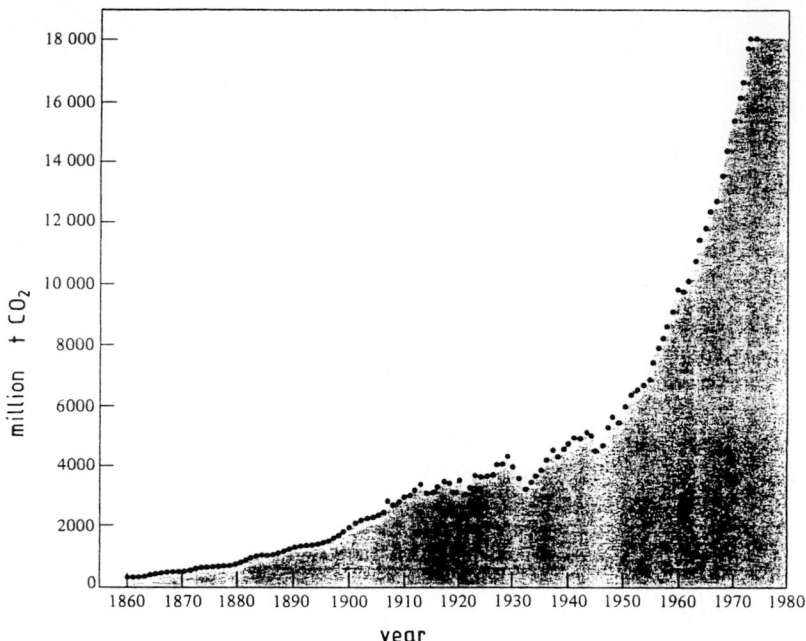

Annual CO_2 production from burning fossil fuels and calcining limestone for the manufacture of cement (from: N. N. Greenwood, A. Earnshaw, *The Chemistry of the Elements*, Pergamon Press, Oxford, New York, Toronto, Sydney, Paris, Frankfurt, **1984**, 303)

Experimental Procedure

The two brass discs are blackened for 2 hours in the potassium peroxodisulfate solution. They are then laid at the bottom of the two beakers, which are standing next to one another on the demonstrating table. The lamp is switched on and the two beakers illuminated for a few minutes. The temperature difference between the beakers is $0 \pm 2\,°C$. One of the beakers is slowly filled with the "greenhouse gas" CO_2. The second beaker is covered with the glass plate while the gas is being passed in. After about 30 seconds the gas inlet tube and the glass plate are removed. The lamp is now switched on again and the two beakers illuminated. Shortly afterwards a temperature difference can be clearly measured: the beaker filled with CO_2 increases its temperature by about $10\,°C$ within a minute. The temperature difference then falls because of convection and diffusion of the carbon dioxide. The blackened brass discs, which re-emit the visible light absorbed in the form of infrared radiation, are vital in this experiment; no temperature difference is generated if they are absent. The determination of the temperature change can be carried out either with a thermocouple or with a normal thermometer. However, in the latter case it is hard to measure the temperatures and no longer possible to compare them directly. The carbon dioxide

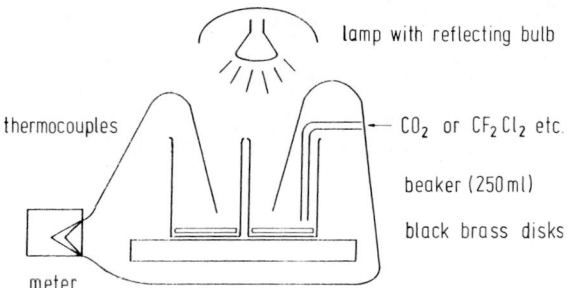

Experimental setup for the simulation of the "greenhouse effect"

concentration in the atmosphere is of course very much lower than in our experiment. The CO_2 could be replaced by CF_2Cl_2 or other chlorofluorocarbons which have a dipole moment.[1] The experimental setup is shown diagrammatically in the figure.

Waste Disposal

The alkaline solution of the oxidising agent can be poured down the drain after being well diluted.

Reference

1 M. Adelhelm, E.-G. Höhn, *J. Chem. Educ.*, **1993**, *70*, 73.

A Barking Dog

To our left, as seen from the entrance, and not far from the porters lodge, was painted on the wall a huge watchdog on a chain, and written above it in square capital letters were the words: Beware of the dog! And my companions actually found it laughable; but I ...

Petronius, Satiricon 29,1

The reaction between carbon disulfide and nitrogen monoxide was carried out *by Liebig* in 1853 during an evening lecture. The audience was so enthusiastic about this experiment that *Liebig* repeated it; this time there was a violent explosion. The Queen of Bavaria was wounded in the cheek. It seems likely that oxygen was used in the second experiment rather than nitrogen monoxide. *Liebig* wrote as follows to *Wöhler* in Göttingen:

As I looked around after the terrible explosion in the room where the audience sat, and saw the blood running from the faces of Queen Therese and Prince Luitpold, my horror was indescribable; I was half dead. Fortunately, the accident had no further unpleasant consequences. Their Majesties behaved in a noble and high-minded way, and all of their concern seemed to focus itself only on me. The Queen sent her personal physician to me that very evening, and every day their Majesties inquired with respect to my health. Old King Ludwig himself came the next day and asked if my wound was serious, and when I said No he exclaimed: Then everything is all right; so long as nothing happened to you, the rest is nothing ...

Safety Precautions

This experiment must always be carried out behind a protective screen. CS_2 has a very low flash point and is highly toxic. The tube must be filled in a well-ventilated hood. Safety glasses and protective gloves must be worn!!

Apparatus

Glass tube 120 cm long and 7–8 cm in diameter closed at one end and fitted with a single-bore stopcock (as for a chromatographic column), rubber stopper, gas burner, 5-mL syringe, stand, clamp, protective shield, protective gloves, safety glasses.

Chemicals

Dinitrogen monoxide N_2O in a pressure cylinder, carbon disulfide CS_2.

Experimental Procedure

N_2O is used rather than NO in this experiment. The glass tube is fixed in an upright position and filled from below with N_2O via the stopcock until all the air has been displaced. 5 mL of CS_2 are then injected into the tube, which is closed with the stopper. The contents of the tube are mixed thoroughly by shaking, so that the CS_2 evaporates and enters the gas phase.

Caution! Excess pressure may be generated! The rubber stopper must be held tightly and the pressure released by opening the stopcock once or twice. The tube is now fixed vertically to the stand with the stoppered end uppermost. The stopper is removed quickly and the gas mixture ignited with the flame of the gas burner.

Explanation

N_2O and CS_2 react to give nitrogen, sulfur, sulfur dioxide and carbon monoxide.

$$3 N_2O + CS_2 \rightarrow 3 N_2 + CO + SO_2 + 1/8 S_8$$

This combustion reaction, just like the well-known hydrogen-oxygen reaction, is accompanied by a characteristic "barking" sound.

A Bromate-Malonic Acid Oscillation Process Catalyzed by Mn^{2+} Ions

Our world is an island, an island of order, an island of physical laws, an island of ideas, an island of trust. We live on our island. Since the world is harmonic, ideas and matter exist in a creative relationship with each other, the world is finite, love is a strong, integrating power in it, we live in a beautiful world!

Perhaps there are other islands – maybe even an entire archipelago. Their order could be different. If it is, we must accept it as equal to ours, since we now recognise the plurality of this world ...[1]

Friedrich Cramer, "Chaos and Order"

An incomparable fascination is the magic of the colorful oscillation reactions which are known today to every chemist, although only a few decades ago they were considered to be completely atypical. Here we shall present an example which, in contrast to the variety of colors often occurring in the course of these processes, demonstrates the principle of a *Belousov-Zhabotinsky* reaction by means of a single color change. Perhaps we can follow *Nietzsche* here: *"There are many things which, once and for all, I do not wish to know. Wisdom also sets limits to knowledge ..."*

Apparatus

1-L beaker, magnetic stirrer with stirrer bar, safety glasses, protective gloves.

Chemicals

Malonic acid $HOOCCH_2COOH$, $KBrO_3$, $MnSO_4 \cdot H_2O$, concentrated H_2SO_4, distilled water.

Experimental Procedure

The beaker contains a cooled mixture of 750 mL of distilled water and 75 mL of concentrated sulfuric acid. 9 g of malonic acid are added, followed on complete dissolution by 8 g of potassium bromate. 1.8 g of manganese sulfate are introduced into this clear solution with constant stirring. The solution first turns orange but after about 80 seconds the color fades. The system is stirred continuously and oscillates between colorless and orange at intervals which can vary between about 20 and 80 seconds. The process does not stop for at least 20 minutes.

Explanation

The *Belousov-Zhabotinsky* reaction demonstrated here is set in train by the reduction of potassium bromate to elemental bromine by malonic acid and manganese(II) sulfate; this is shown by the orange coloration. The reaction of the bromine with malonic acid to give mono or dibromomalonic acid leads to decolorisation. At the same time more bromine is formed in the initial redox process, and this again replaces one or two hydrogen atoms of the malonic acid. The process is repeated many times; the start reaction is inhibited by complexation of the brominated malonic acid by Mn(II) ions, so that the oscillation slowly comes to an end.[2]

Waste Disposal

The reaction mixture is treated with soda solution and reduced to a small volume; this residue is transferred to the container used for collecting less toxic inorganic salts.

References

1 F. Cramer, *"Chaos and Order"*, VCH Weinheim, **1993**, 232.
2 P. Glansdorff, I. Prigogine, *Thermodynamic Theory of Structure, Stability and Fluctuations*, Wiley-Interscience, New York, **1971**.

A Green-Blue-Red Belousov-Zhabotinsky Reaction

Kurt Seligmann describes a special type of oscillation in his *"Weltreich der Magie"*.

According to Peter von Lothringen, *Abbot of Vallemont (1649–1721), the entire secret of the vampire consists of salt, heat, and motion. As for the rest, everything that existed once before can appear again: there is nothing alarming about that. The dead can come back to life, at least temporarily, just as plants and animals can be revived. One places the germ of life from the seeds of a beautiful rose in a flask. The germ is then combusted to an ash and mixed with dew, specifically with the amount required for a distillation. One subsequently separates the salt from the ash, mixes it with the distilled dew, and seals the flask with powdered glass and borax. The vessel is then left for a month on fresh horse manure, and later exposed alternately to sunlight and moonlight. If the gelatinous mass on the bottom of the flask swells up, one can be sure that the experiment is a success. As soon as the vessel is exposed to the sun, an image of the rose will appear therein in all its beauty, together with leaves and buds. It vanishes upon cooling, and reappears upon warming. This process can be repeated as often as desired.*

Apparatus

Three 600-mL beakers, 2-L beaker or glass jar, magnetic stirrer with stirrer bar, 500-mL and 50-mL measuring cylinders, safety glasses, protective gloves.

Chemicals

$KBrO_3$ or $NaBrO_3$, distilled water, malonic acid, KBr or $NaBr$, $Ce(NH_4)_2(NO_3)_6$, 2.7 mol/L H_2SO_4 (150 mL concentrated H_2SO_4 diluted to give 1 L of solution), 0.5 % ferroin solution (0.23 g $FeSO_4 \cdot 7 H_2O$ and 0.56 g 1,10-phenanthroline $C_{12}H_8N_2$ dissolved in 100 mL of water).

Solution A

19 g $KBrO_3$ (17 g $NaBrO_3$) are dissolved in 500 mL of distilled water.

Solution B

16 g of malonic acid and 3.5 g KBr (3.0 g NaBr) are dissolved in 500 mL of distilled water.

Solution C

5.3 g $Ce(NH_4)_2(NO_3)_6$ are dissolved in 500 mL of 2.7 mol/L H_2SO_4.

Experimental Procedure

The large beaker (or glass jar) is placed on the magnetic stirrer; the solutions A and B are added with stirring. After about a minute the solution C and 30 mL 0.5 % ferroin solution are added to the colorless mixture. The solution becomes a cloudy green. After about a minute the oscillation begins: the green solution turns blue, violet and then red, before finally returning to green. The oscillation comes to an end after about 20 minutes.[1]

Explanation

The addition of the redox indicator ferroin causes the "classical" oscillation system, which was first described in detail by *A. M. Zhabotinsky*,[2] to go through the various differently colored stationary states, which undergo interconversion at regular intervals.

Waste Disposal

The solution is rendered slightly alkaline, the volume of liquid reduced and the precipitate separated off, the latter is transferred to the container used for less toxic inorganic salts, while the filtrate is washed down the drain.

References

1   B. Z. Shakhashiri, *Chemical Demonstrations, A Handbook for Teachers of Chemistry,* University of Wisconsin Press, Madison, London, **1983,** 2, 257.
2   A. M. Zhabotinsky, *Biofizika* (Russian), **1964,** 9, 306.

An Oscillating Platinum Wire

To encounter him, what need I more,
Than to utter the spell of the FOUR!
SALAMANDER shall glow,
UNDINA shall flow,
SYLPH melt in the Sky,
On Earth KOBOLD ply!

The Elements who doth not known,
Their energies,
Their properties,
The Spirits' Master
ne'er will grow!

In flames disappear,
Salamander!
In a torrent, rush clear,
Undina!
Shine a Meteor fair,
Thou Sylphid!
My house be thy care,
Incubus, Incubus!
Step forth, and finish the charm for us!

Johann Wolfgang von Goethe, "Faust, The First Part",
Faust in his Study

Apparatus

500-mL Erlenmeyer flask, gas burner, cigarette lighter, fire-resistant support, safety glasses, protective gloves.

Chemicals

Methanol, platinum wire (thickness 0.8–1.0 mm), nickel wire about 0.5 mm thick.

Experimental Procedure

80 mL of methanol are placed in the wide-necked flask and heated to between 40 and 60 °C. A small figure formed from the platinum wire is now hung (using the nickel wire) about 1 cm above the surface of the methanol. The figure must be heated with the gas burner or a cigarette lighter prior to the experiment. When the room is darkened it can be seen that the platinum wire begins to glow; when the glow reaches its peak the methanol catches fire, burns for a short while and goes out again. At this point the wire glows only faintly, but the glow rapidly increases in intensity and again the methanol catches fire. This cycle continues until the methanol has been used up. The effect can still be observed even if the platinum figure is as much as 5 cm above the surface of the methanol.[1]

Explanation

The catalytic oxidation of methanol produces formaldehyde and water (eqn. 1):

(1) $CH_3OH + 1/2\ O_2 \rightarrow CH_2O + H_2O$

The combustion of methanol produces carbon dioxide and water (eqn. 2):

(2) $CH_3OH + 3/2\ O_2 \rightarrow CO_2 + 2\ H_2O$

Waste Disposal

The remaining methanol is placed in the container used for collecting non-halogenated organic solvents.

Reference

1 D. L. Coffing, J. L. Wile, *J. Chem. Educ.*, **1993**, *70*, 585.

Green-Red-Yellow:
An Unusual Traffic Light

Finally, there is no existence that is constant, either of our being or of that of objects. And we, and our judgment, and all mortal things go on flowing and rolling unceasingly. Thus nothing certain can be established about one thing by another, both the judging and the judged being in continual change and motion.

We have no communication with being, because every human nature is always midway between birth and death, offering only a dim semblance and shadow of itself, and an uncertain and feeble opinion.

Michel de Montaigne
Essay "Apologie des Raimund Sebundus"

Apparatus

2-L beaker, 1-L beaker, 250-mL beaker, gas burner, thermometer, safety glasses, protective gloves.

Chemicals

Solution A: 14 g glucose in 700 mL of water
Solution B: 6 g NaOH in 200 mL of water
0.04 g indigo carmine (the disodium salt of indigo-5,5′-disulfonic acid).

Experimental Procedure

The 1-L beaker containing the glucose solution is warmed to about 35 °C. 0.04 g of the indicator indigo carmine is added, and the solution turns blue. Solution B is added to give a green color. After a short time there is a color change via red to golden yellow. If this golden yellow solution is now suddenly poured from a height of at least 60 cm into the empty two-liter beaker it turns back to green, but the color again changes to red and then to golden-yellow. The demonstration can be repeated several times, the colors becoming a little paler each time.[1]

Explanation

The blue dyestuff indigo carmine (used for wool) is extremely air-sensitive, so that the oxygen in the air causes the color to return to green when the reduced dye solution is poured into the empty beaker.

Waste Disposal

The solution can be poured down the drain.

Reference

1 B. Iddon, *Chemistry in Britain,* **1993,** *29,* 657.

A Flashing Blue Light

The speed with which a chamaeleon can change its color is proverbial. The species whose members undergo the fastest color changes and show the greatest variety of colors is the octopus family; here the color change is brought about by means of specialised cells which contain pigments known as chromatophores. Each of these chromatophores is surrounded by an elastic membrane, the size of which can be controlled by radial muscles. When these contract the membrane is forced apart to give a flat disc, so that the pigment is now uniformly distributed over a large area. This pigment is pressed into a small sphere by the membrane when the muscles relax. An octopus or cuttlefish can undergo a continuous change of color by extending or contracting the required chromatophores; this process is remarkably fast, and no other species can compete with an octopus in the rate at which it can change color.[1]

Apparatus

Magnetic stirrer, 500-mL glass jar, three 200-mL beakers, magnetic bar, safety glasses, protective gloves.

Chemicals

Sodium iodate $NaIO_3$, 1 mol/L H_2SO_4, malonic acid $HOOCCH_2COOH$, manganese(II) sulfate $MnSO_4 \cdot H_2O$, soluble starch, 10 % H_2O_2 solution, distilled water.

Solution A

1.5 g sodium iodate and 10 mL 1 mol/L H_2SO_4 in 100 mL of distilled water.

Solution B

1 g malonic acid, 1.5 g manganese sulfate and 10 mL 1 % starch solution in 100 mL of distilled water.

Solution C

135 mL 10 % H_2O_2.

Experimental Procedure

The three colorless solutions are poured simultaneously into the glass jar and stirred well. After a short time the oscillation between blue and colorless starts.

Explanation

The oscillation process, the theoretical principles of which are described in Experiment 96, leads to the periodic formation of free iodine, which is characterised by the blue color of the starch-iodine complex.[2]

Waste Disposal

The reaction mixture is stirred with milk of lime. It is decanted and the solid residue transferred to the container used for collecting less toxic inorganic waste. The clear solution is brought to a pH of 7 to 8 and poured down the drain.

References

1 M. and P. Fogden, *Farbe and Verhalten im Tierreich*, Herder-Verlag, Freiburg, **1975**.
2 K. Häusler, H. Rampf, R. Reichelt, *Experimente für den Chemieunterricht*, R. Oldenburg Verlag München, **1991**, 342.

The Döbereiner Cigarette Lighter: Physical and Chemical Properties of Hydrogen

Fire has played an essential part in the development of human civilisation and culture. Although it was known from experience that air was required to promote and sustain combustion, oxygen was not discovered until 1774 (by *Priestley*). In 1776 *Cavendish* discovered hydrogen. *Lavoisier* showed that water was decomposed by iron filings, that the amount of hydrogen formed corresponded to the amount of oxygen given up to the water by the iron filings, and that water was formed when hydrogen was burned. Thus *Lavoisier* had shown experimentally that water is composed of oxygen and hydrogen, not only synthetically by burning the gases together, but also analytically by studying the decomposition of water.

In the year 1782 both *Cavallo* and *Lichtenberg* in Göttingen prepared soap bubbles filled with hydrogen. They observed that these rose in the air and burst on reaching the ceiling.

Hydrogen and oxygen combine quantitatively at room temperature in the presence of finely divided palladium or platinum to produce water. In 1823 *Döbereiner* used this catalytic reaction to construct a cigarette lighter.

In his publication of 1823, written in Jena and entitled *"Über neu entdeckte höchst merkwürdige Eigenschaften des Platins"* (*"On newly discovered, highly curious properties of platinum"*) *Döbereiner* wrote as follows:

> *The fire-inducing activity described in the last experiment but one, observed when detonating gas [a hydrogen and oxygen mixture] is brought in contact with platinum, led me to the idea of using this to devise a new type of igniting device or night lamp, etc.*
>
> *I carried out a large number of experiments in order to establish the conditions under which platinum would be brought to glowing with the minimum application of hydrogen gas, and eventually determined that the desired phenomenon occurs most splendidly if one allows the hydrogen gas to stream out of a gas reservoir (a so-called electrical lighter) and onto spongy platinum powder through a glass capillary tube that has been bent*

downward, where the platinum is contained in a watch glass or near the pointed end of a sealed glass funnel. In particular, the gas stream should mix with atmospheric air prior to making contact with the platinum (which transpires if the tip of the capillary is located 1, 3/2, to 2 inches above the platinum). The platinum powder then becomes almost instantaneously red- and then white-hot, and remains so as long as the hydrogen gas continues to flow. With a strong flow, the hydrogen ignites.

Döbereiner became Professor of Chemistry, Pharmacy, and Technology at Jena in 1810. He was also *Goethe's* adviser in chemical matters. He complained to the latter in a letter of 1812 regarding his poor working conditions:

Investigation is made doubly difficult for me by the fact that I am unable to carry out any experimental work myself in my rented apartment, being forced instead to do it all in the Ducal laboratory, where in winter the warmest chemist would be benumbed within a few hours from the cold. If I could only, perhaps through the grace of the illustrious Duke, acquire a residence ... in which I could arrange things entirely in concert with my scientific goals and needs, then I would devote all of my time to chemical investigations, and let no unexplored area of science escape my attention.

Safety Measures

Before igniting the hydrogen flame the gas formed should be allowed to escape for about a minute in order to make sure that oxygen gas is no longer present in the reaction tube. Safety glasses should be worn!

Apparatus

Döbereiner cigarette lighter, safety glasses, protective gloves.
The Döbereiner cigarette lighter (see figure) consists of a storage jar (about 30 cm in height and 10 cm in diameter) for the acid and a lid to which is attached a reaction tube open at the bottom; this serves to collect the hydrogen gas formed, which can be taken off via a nozzle with a gas tap. The reaction tube also contains a reaction dish in which the zinc granules are placed. A holder for a piece of platinum sponge is situated about 5 cm from the gas tap.

Chemicals

Zinc granules (about 20–25 g), 6 mol/L hydrochloric acid (370–400 mL; the hydrochloric acid can be replaced by sulfuric acid), platinum sponge. The hydrogen flame, which is normally not easily visible, is colored yellow if a few crystals of sodium chloride are added to the acid.

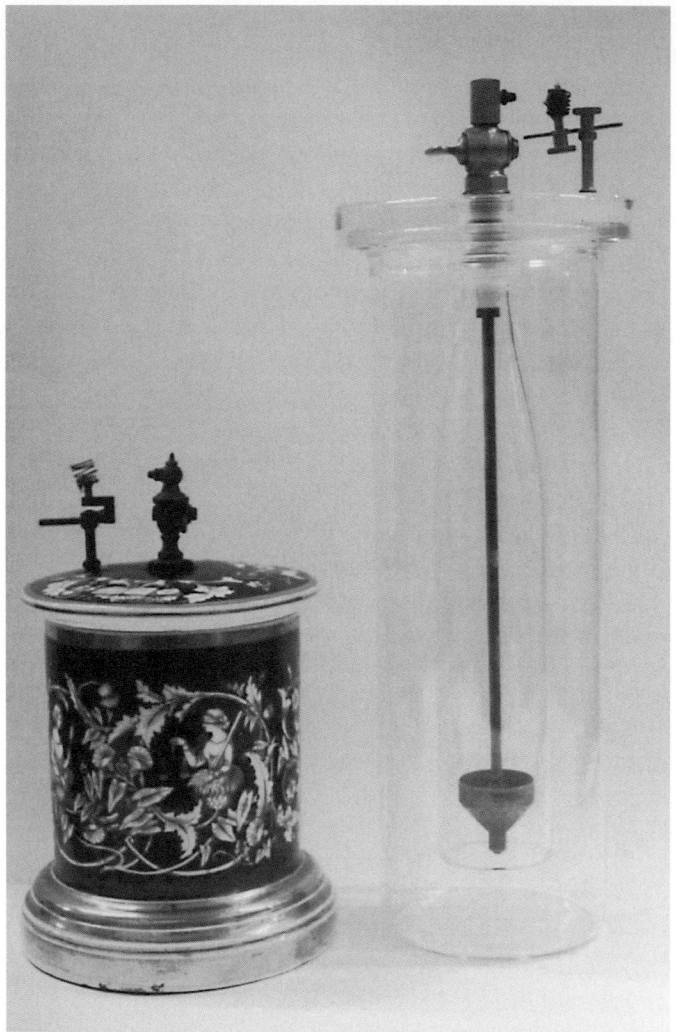

Original Döbereiner cigarette lighter and a modern version made of glass

Experimental Procedure

The required amount of acid is placed in the storage jar and the zinc placed in the reaction dish. The lid is placed on the storage jar before the experiment is carried out. The bell-shaped reaction tube with the reaction dish is now immersed in the acid to a depth of about 2 cm. A vigorous evolution of gas is at once observed. If the gas tap is opened, the gas can escape and is ignited at the platinum sponge. The hydrogen flame burns only for as long as gas is allowed to escape, and goes out at once when the gas tap is closed. It can be seen that the gas then formed pushes down the acid in the reaction tube until it is no longer in contact with the metal.

Waste Disposal

The acidic solution containing zinc salts is transferred to the container used for collecting less toxic inorganic waste.

The Landolt Experiment

I pass with relief from the tossing sea of Cause and Theory to the firm ground of Result and Fact.

Sir Winston S. Churchill (1874–1965),
"The Malakand Field Force"

In 1886 *Hans Landolt (1831–1910)* published a pioneering publication in the *Berichte der deutschen chemischen Gesellschaft* entitled "*Ueber die Zeitdauer der Reaction zwischen Jodsäure und schwefliger Säure*" ("On the duration of the reaction between iodic acid and sulforous acid"):[1]

If excess iodic acid solution is added to aqueous sulfurous acid it is well known that iodine separates from the mixture. The reaction occurs immediately if the liquids are concentrated; however, if the same liquids are used in dilute form it leads to the remarkable phenomenon that such a mixture treated with a little starch initially remains completely clear, and only after the passage of a certain amount of time suddenly becomes blue, which may require a few seconds up to minutes. Using the same amounts of the two solutions and maintaining a specific temperature, the time interval from the moment of mixing to the appearance of the blue color is entirely constant, with a value that can easily be determined with a clock.

Apparatus

Three 1000-mL beakers, six 250-mL beakers, three 50-mL beakers, three glass rods, 500-mL measuring cylinder, stopclock, safety glasses, protective gloves.

Chemicals

Distilled water, KIO_3, Na_2SO_3, starch, concentrated H_2SO_4, ethanol.

Solution A

8.6 g KIO_3 dissolved in 200 mL of water.

Solution B

8 g of concentrated H_2SO_4, 20 mL of ethanol and 2.32 g Na_2SO_3 dissolved in 2000 mL of water.

Solution C

1 g of starch dissolved in 500 mL of hot water; must be freshly prepared.

Experimental Procedure

Distilled water, starch solution (C), and sulfite solution (B) (for amounts see below) are added one after the other to the three beakers; after (B) has been added the mixtures are stirred well. The required amount of iodate solution (A) is added last and must be poured simultaneously into all three beakers:

Beaker 1: 200 mL H_2O + 20 mL of solution (C) + 100 mL of solution (B) + 100 mL of solution (A)

Beaker 2: 400 mL H_2O + 20 mL of solution (C) + 100 mL of solution (B) + 100 mL of solution (A)

Beaker 3: 600 mL H_2O + 20 mL of solution (C) + 100 mL of solution (B) + 100 mL of solution (A)

After solution (A) has been added the contents of all three beakers are stirred for about 7 seconds, and left to stand. The color change to blue takes place in beaker 1 after 12 to 15 seconds, in beaker 2 after 25 to 30 seconds and in beaker 3 after about 1 minute. It is best to use a stopclock with a large face and a large second hand to measure these times.

Detailed studies on the Landolt reaction and on its use for lecture demonstrations were carried out between 1917 and 1922 by *J. Eggert* and *A. Skrabal*. The reaction mechanism presented below was confirmed. *Skrabal* writes as follows.[2]

Many lecture experiments have been proposed for demonstrating the laws of chemical kinetics. Nevertheless, they fail to meet those criteria one can justifiably impose for lecture-demonstration purposes from the standpoint of simplicity, transparency, impressiveness, and reliability. With respect to impressiveness and clarity of the phenomenon, as well as the element of surprise, the Landolt experiment stands out in first place. Conducted in the usual way, however, it is necessarily deprived of its simplicity and transparency.

The reaction to be demonstrated should be a simple one, and it should be based on a simple kinetic expression. Moreover, it should proceed with a constant velocity. In such a case the reaction velocity need not be expressed as a differential quotient, but rather as the straightforward quotient:

$$\text{Reaction velocity} = \frac{\text{Length of the reaction path}}{\text{Time of the reaction}}$$

Just as with a speeding railroad train or a running horse, the rate can be established on the basis of a particular segment of a journey and the time required to complete that segment. Constant velocity is given if the measured segment is small relative to the entire journey. The reaction time is a function of the length of the reaction segment and the constant reaction velocity, and is given by the following equation:

$$\text{Reaction time} = \frac{\text{Length of the reaction segment}}{\text{Reaction velocity}}$$

All these requirements can be realised with the Landolt reaction.

Explanation

The hydrogen sulfite reduces the iodate solution to iodide:

$$IO_3^- + 3\ HSO_3^- \xrightarrow{\text{slow}} I^- + 3\ HSO_4^-$$

Iodide and iodate react under acid conditions to give elemental iodine, which forms a blue inclusion complex with starch:

$$5\ I^- + IO_3^- + 6\ H^+ \xrightarrow{\text{fast}} 3\ I_2 + 3\ H_2O$$

Iodine is however also reduced very rapidly by hydrogen sulfite ions to iodide ions:

$$I_2 + HSO_3^- + 3\ H_2O \xrightarrow{\text{very fast}} 2\ I^- + HSO_4^- + 2\ H_3O^+$$

Thus the blue inclusion complex becomes visible only when all the hydrogen sulfite has been consumed. According to studies by *R. C. Teitelbaum, S. L. Ruby* and *T. J. Marks*[3] the blue inclusion compound consists of the amylose component of the starch and the polyhalogenide anion I_5^-, this was established by comparing the Raman and ^{129}I Mößbauer spectra of the blue-black amylose-iodine complex with those of the adduct between trimesic acid hydrate and $H^+I_5^-$, the structure of which is known (see figures).

   The discovery of the starch-iodine reaction is ascribed to the French chemists *J. J. Colin* and *H. F. Gaultier de Claubry* (1814). The chemist *F. Stromeyer* from Göttingen studied this reaction independently, writing as follows on January 15th to Professor *Gilbert*, the editor of the *Annalen der Physik*:

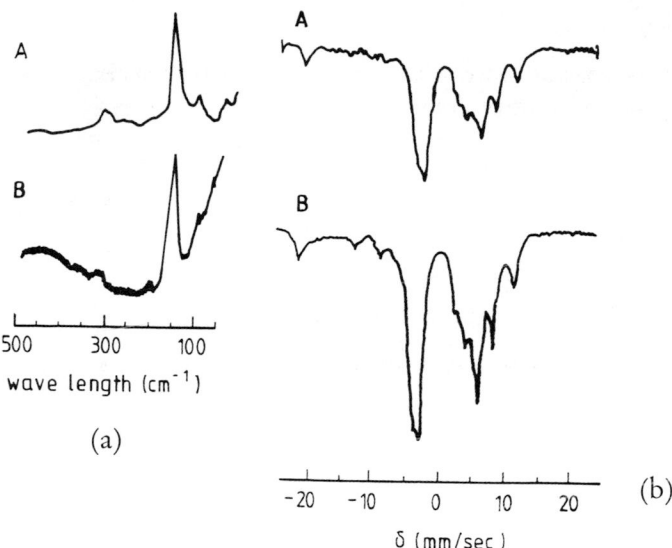

Resonance Raman spectra (a) and ^{129}I Mößbauer spectra (b) of the starch-iodine complex (A) and the model substance (trimesic acid $H_2O)_{10} \cdot HI_5$ (B)

Not only vapors of iodine, but also solutions of this material in water, alcohol, and anesthetic ether cause starch to be colored a magnificent indigo blue. As I further pursued this observation, which I had made in the course of my experiments with iodine, I quickly became persuaded that this singular behavior placed in our hands a means for easily and reliably detecting even the smallest amount of iodine. Based on H. Davys experiences, polished silver is the best reagent for iodine among those so far investigated, and I have convinced myself through my own experiments of the extraordinary sensitivity of silver. Only starch (amidone) exceeds the latter by a wise margin in terms of the rapidity of the effect, without falling behind the same in intensity. Starch through its coloration reveals 1/200 000 to 1/250 000 part of iodine on the spot, and with amounts corresponding to only 1/350 000 to 1/450 000 the coloration still occurs within a few minutes. By contrast, the effect of polished silver already ceases to be instantaneous with quantities of 1/25 000 part of iodine, and shows initial blackening only after 1/4 of an hour; with amounts of 1/100 000 to 1/150 000 of iodine this blackening begins only after 1 or 3/2 hours; and in liquids that contain only 1/350 000 to 1/450 000 part of iodine the discoloration of the silver first becomes noticeable after 18 to 24 hours.

Waste Disposal

The solutions contain only small amounts of harmless substances, so that they can be poured down the drain.

References

1 H. Landolt, *Ber. Dtsch. Chem. Ges.*, **1886**, *19*, 1317.
2 A. Skrabal, *Z. Elektrochemie*, **1922**, *28*, 224.
3 R. C. Teitelbaum, S. L. Ruby, T. J. Marks, *J. Am. Chem. Soc.*, **1978**, *100*, 3215.

103

"Home-Brewed Beer"

For this reason alchemy is a chaste harlot with many lovers, all of whom are disappointed, however, and none is granted her embrace. She transforms the stupid into imbeciles, the rich into beggars, the philosophers into babblers, and the deceived into elegant deceivers.

Trithemius, "Annalium Hirsaugensium Tomi II" (1690)

Emmer, a type of wheat which grows wild in the Near East, was used by the Egyptians for making bread and beer at the time of the Pharaohs, as it is shown by excavations in the ancient Egyptian city of Tell El-Amarna. Here ruins of a large bakery were found in which beer was also brewed. This factory was built at the order of the Pharaoh *Amenophis IV*, also known as *Echnaton*, who ruled from 1364 to 1348 B.C. *Echnaton* was the husband of *Nefertiti* and the father-in-law of the Pharaoh *Tutenkhamen*, whose almost undamaged grave was discovered in the year 1922 with all its treasures. There archaeologists found molds for baking bread as well as broken vessels for brewing beer.

The vessels which were found in Tell El-Amarna had a volume of between 40 and 50 liters. Wall paintings showing brewers and their tools, which were discovered in the ancient Egyptian temples in Luxor and Thebes, provided Egyptologists with further evidence. Studies of wheat grains showed that the Egyptians were able to produce malt at the time of the Pharaohs. The wall paintings and inscriptions indicate that they soaked wheat grains until these germinated and then crushed them to give a paste to which yeast was added. The sourdough thus formed was baked for a short time, crumbled and allowed to ferment with water in large vessels.

Nobody knows how the Egyptians were able to keep the fermenting liquid cool. Too much heat destroys the enzymes which are necessary for the formation of beer. However, the vessels used for brewing were porous, which could have made evaporation possible; this would in turn have led to cooling.

Hops were probably not used by the ancient Egyptians for their beer, which was more likely brewed using herbs, cinnamon and fruit.

Apparatus

Pint or 0.4-L beer glass, two 250-mL beakers, two 100-mL measuring cylinders.

Chemicals

Solutions A and B from the Landolt experiment (102), washing-up liquid (or similar).

Experimental Procedure

The solutions A and B from the previous **Landolt experiment** are used. 100 mL of the sulfite solution (B) and of the iodate solution (A) are each diluted with 100 mL of distilled water. 2 mL of the washing-up liquid are placed in the beer glass prior to the experiment.

Both solutions are now poured simultaneously into the beer glass. Because of the presence of the washing-up liquid a clear frothy solution is formed, which turns yellowish-brown after about 10 seconds, so that "beer" now appears to have been formed (colored figure 30).

Under no account may this solution be drunk!

Waste Disposal

The solutions contain only small amounts of harmless substances and can be poured down the drain.

104

"Artificial Coke"

Our concern is a secret within a secret, a secret of something that remains veiled, a secret that only another secret can explain, a secrete about a secret that can be answered with a secret.

Ga'far al-Sadiq, Sixth Imam

... Pemberton?

Quite right: *John Pemberton*, the discoverer of the world's second favorite drink (after water), *Coca-Cola*. The man is long dead, but his drink lives on. In 1886 he received $ 2300 for the patent of his discovery, which he described as "the ideal tonic for hangovers". The recipe for the one and only *Coca-Cola* is one of the world's best-kept secrets!

Apparatus

Magnetic stirrer with magnetic bar, empty liter Coca-Cola bottle, two 250-mL beakers, 20-mL measuring cylinder, funnel, safety glasses, protective gloves.

Chemicals

Solution 1 0.2 % starch solution
Solution 2 5 g iodic acid in 100 mL of water
Solution 3 2.1 g Na_2SO_3 in 100 mL of water.

Experimental Procedure

The empty liter Coca-Cola bottle is stood on the magnetic stirrer. 930 mL of distilled water are poured in, followed by 15 mL of the 0.2 % starch solution. 20 mL of the iodic acid solution are now poured in with stirring, followed by 20 mL of the sulfite solution. After about 30 seconds the solution turns to a typical Coca-Cola-like dark brown (see colored figure 31).

Under no circumstances may this solution be drunk!

Explanation

The combination of the blue color of the starch-iodine complex and excess iodine gives a dark brown solution.

Waste Disposal

The brown solution may be poured down the drain.

105

The Chlorine-Hydrogen Reaction

Physicists and chemists are justified in ascribing most of the success in their work to their research methods. Every chemical or physical investigation that more or less bears the stamp of completion can be described in results comprising but a few words. These few words alone represent eternal truths the discovery of which required innumerable experiments and questions; the research projects themselves, the tedious experiments and the complex apparatus, are soon forgotten once the truth has been established; they are the ladders, the shafts, and the tools that were indispensable in reaching the precious ore; they are the props and the air shafts that keep the mines free from water and dampness.

Georg Christoph Lichtenberg

Letter of *Robert Wilhelm Bunsen* to *Sir Henry Enfield Roscoe:*

I am very pleased, my dearest friend, that you plan to come over here early in the spring so that we can bring our work to a conclusion. I have in recent days carried out the demonstration experiment we discussed here, and I hope with the help of the same to acquire some rather interesting data for our investigation. The experiment is most successful when conducted as follows:

The amount of gas produced in our hydrogen chlorine generator is led in the dark through the system of glass bulbs B, which are about the size of a doves egg and blown so thin *that they can be crushed* with the finger, *and prior to passage of the gas have been thoroughly wetted inside with water. Once the apparatus is full the rubber ligatures are tied together at a and cut cleanly at b, and the rubber ends are dipped into molten yellow wax. These bulbs can be kept several days without spoiling, and if one of them is held near an open window it explodes instantaneously. The bulbs can be held in the hand quite safely by their stems provided one puts on a glove and holds a small glass plate between the bulb and the face. The explosion occurs on the greyest of days, even with rather heavy fog, and it is scarcely more dangerous than the ignition of a soap bubble filled with O and H. I have now had a red and a blue pane of glass installed in the auditorium window: if a bulb is held in front of the former it remains unchanged, whereas in front of the latter it explodes instantaneously. There is scarcely another lecture demonstration that could be more beautiful than this.*

... with respect to our inductive effects in light of various colors. I want to investigate next whether the bulbs can be induced in red light, or whether a bulb induced in blue light acquires the ability to explode in red, which seems to me very probably given the continuing effect of red light with Daguerreotype plates. Repeat the experiment and demonstrate it on some appropriate occasion at the Chemical Society. *I send my cordial greetings to* Williamson Graham *and Mr.* Russel, *whose place I have reserved.*

Safety Precautions

Chlorine is highly toxic, and can cause lung damage (bronchitis) on inhalation even at low concentrations. Hospital treatment is necessary if larger amounts of chlorine are inhaled, as a pulmonary embolism can result. The mixture of chlorine and hydrogen must be prepared in a darkened room. Safety glasses and protective gloves must be worn.

Apparatus

Plexiglass box (30 × 30 × 30 cm with walls 5 mm thick, open at the bottom) with a fixing device for the test tube, large test tube with rubber stopper, flash lamp, small glass or plastic trough, measuring cylinder, pane of red glass, safety glasses, protective gloves.

Chemicals

Hydrogen, chlorine (both from pressure cylinders).

Experimental Procedure

The volume of the test tube is measured prior to the experiment using water and a measuring cylinder and the required volumes of chlorine and hydrogen are marked on its side. The room must be darkened before the chlorine-hydrogen mixture is prepared. The test tube is filled with water and laid in the trough, which is filled with water. Chlorine and hydrogen are then passed into the test tube up to the corresponding marks via gas inlet tubes. The test tube is now closed with a rubber stopper and placed in the plexiglass box, which is in turn placed on the demonstrating table so that the latter forms the "lid" which closes it completely: the plexiglass box thus functions as a protective cover on all sides. The pane of red glass is placed between the flash lamp and the box. No reaction occurs when the flash lamp is set off. The glass is removed and the flash lamp set off again. The test tube explodes with a loud bang.

Explanation

This is a free radical chain reaction which is initiated photochemically by the homolytic cleavage of chlorine molecules to give chlorine atoms (243.5 kJ + Cl_2 = 2 Cl˙). The cleavage of one mole of chlorine by photodissociation requires 1 mole of light quanta (hv). Since an amount of energy equal to 243.5 kJ corresponds to a wavelength of 491 nm*, only blue light or light of even shorter wavelengths can be used, but not longer-wavelength yellow or red light. Thus the chlorine-hydrogen reaction is initiated by blue light but not by red light.[1]

Reference

1 O. Krätz, H. J. Bersch, *Chem. Exp. Didaktik*, **1975**, *1*, 63.

---

* It follows from $h v N_L = h \times 1/\lambda \times c \times N_L = (6.62 \times 10^{-34})(1/\lambda)(3 \times 10^8)(6.023 \times 10^{23}) = 243.5 \times 10^3$ that $\lambda = 4.912 \times 10^{-7}$ m, i.e. ≈ 491 nm

106

Big Bang in a Tin Can

Faust asks:

> But from this house how may we get away
> Where is thy coach, thy horses, servants, where?

and *Mephistopheles* replies:

> This mantle we have only to display,
> It soon shall waft us thro' the air.
> For the bold trip we are about to take
> You must no heavy bundle make.
> An ounce of gas, if duly rarified,
> (Which I at pleasure can provide,)
> Will soon convey us out of sight
> Of this dull spot.

Johann Wolfgang von Goethe, "Faust, The First Part", Faust's Study

A plaque in the Jardin des Tuileries in Paris reminds us that the physicists *Charles* and *Robert* were the first to fill a balloon with flammable air (hydrogen) in 1783.

On the evening of the 6th of May 1937 the German airship Hindenburg went up in flames during its landing maneuver on the Lakehurst navy air force base. This airship, the largest and most luxurious ever built, was filled with hydrogen, the lightest of all chemical elements. Today it is standard procedure to fill the fuel tanks of space rockets with liquid hydrogen and oxygen. The combustion of these two elements, which occurs at around 2000 °C, provides a maximum thrust for the rockets.

When *Henry Cavendish* discovered hydrogen in 1766, the phlogiston theory, first advanced by the German chemist and physician *Georg Ernst Stahl* (1659–1734) was still accepted. In 1784 *Cavendish* had demonstrated that hydrogen combines with oxygen to give water, but he still believed that he had found the elusive phlogiston in the course of this reaction.

We know today that oxidation reactions take place during combustion with oxygen, and with much goodwill "phlogiston" can be interpreted as the energy set free during these reactions.

Safety Precautions

Hydrogen and air form explosive mixtures. Safety glasses must be worn!

Apparatus

Large tin can (volume approx. 1 liter), cigarette lighter, gas tubing, safety glasses, protective gloves.

Chemicals

Hydrogen.

Experimental Procedure

A small hole (about 2 mm in diameter) is made in the base of the can with the help of a nail. The can is placed on the table with its open top facing downwards but is tilted, for example by laying a glass rod under one side. Hydrogen is passed into the can through a piece of flexible tubing until it is absolutely certain that no air is left in the can. The hydrogen is turned off and the tubing removed. The hydrogen which escapes from the can through the small hole is then ignited with the cigarette lighter. After about 30 seconds there is a very loud bang!

Explanation

The hydrogen burns with an almost invisible flame, thus drawing air into the can at the base. The hydrogen left in the can forms an explosive mixture with the air, and this mixture is ignited by the flame. The hydrogen reacts with the oxygen in the air to form water.

Gas Explosions

Me thinks the Chymists, in their searches after the truth, are not unlike the navigators of Solomon's Tarshish fleet, who brought home from their tedious voyages not only gold, and silver and ivory, but apes and peacocks too: for so the writings of several of your hermetick philosophers present is, together with divers substantial and noble experiments, theories, which either like peacock feathers make a great show, but neither solid nor useful or else like apes; if they have some appearance of being rational, are blemished with some absurdity or other, that, when they are attentively considered, make them appear ridiculous.

Robert Boyle (1626–1691), "The Sceptical Chymist"

Serious accidents due to explosions often occur because of a series of unfortunate coincidences. Thus during a lecture at a university the well-known experiments involving oxygen were to be carried out in large 5 liter flasks. The lecturer's assistant noticed that one of the flasks was wet and asked an apprentice to dry it. This he did by rinsing the flask with acetone, but he failed to remove the remaining acetone from the flask. The assistant filled the flask with oxygen and used a glowing wood splint to check the contents. A violent explosion shattered the flask, and splitters of glass shot out in all directions. Fortunately this happened during the preparations for the lecture and not during the lecture itself!

Apparatus

Narrow-necked polythene bottles (50 or 100-mL), piezoelectric spark generator, sparking plug attached to base plate, gas inlet tubes, 1-L beaker, dropping pipette, measuring cylinder, safety glasses, protective gloves.

Chemicals

Hydrogen, oxygen, light petroleum, water.

Experimental Procedure[1]

(I): A 50-mL polythene bottle is filled with oxygen until all the air has been displaced and closed with a stopper. Two drops of light petroleum are

Experimental setup for detonation of gas mixtures

placed in the bottle using a dropping pipette; the bottle is shaken well and pushed over the thread of the sparking plug (see figure). The mixture is ignited by means of the piezoelectric spark generator. There is a loud bang and the polythene bottle is generally torn open along one side; sometimes it can even fly into the air.

(II): Prior to the experiment the amounts of O_2 and H_2 to be filled into the 100-mL polythene bottle are determined using a measuring cylinder and appropriate marks made on the bottle. The oxyhydrogen gas is prepared by filling the bottle with water and placing it in the beaker, which is also full of water. Oxygen is passed in until the appropriate mark is reached, followed by the necessary amount of hydrogen. The bottle is closed with a stopper and shaken. It is then pushed onto the sparking plug and the gas ignited. There is again a loud bang, and the bottle flies into the air.

Reference

1 M. Jäckel, H. Willner, *Chemie in unserer Zeit*, **1989**, *23*, 64.

108

The Reaction of a Mixture
of Acetylene and Air

*Vigorous evolution of gas, quick coloration to brown, and the formation of
precipitates; there, hidden, was the treasure of possibility in the bubbles of
foam on the surface, which were observed in the reaction vessel in a corner
of our small laboratory! For organic chemists, facing such an unpredictable
phenomenon is not uncommon. In flasks, that which can never be predicted
by thought or discussion with co-workers often happens.*

Teruaki Mukaiyama, "Challenges in Synthetic Organic Chemistry"

The discovery of acetylene (ethyne) by *Davy* formed the basis of industrial
organic chemistry for many decades; only much later was it replaced by petro-
chemistry.

In 1839 *E. Davy* presented a report in the *Transactions of the Royal Irish Academy* entitled "On a new gaseous compound of carbon and hydrogen":

*In attempting to make potassium, on a large scale, in an iron bottle, by
what has been called* Brunner's *method, i.e. by strongly heating a mixture
of previously calcined cream of tartar, and about 1/14 of dry charcoal powder, I failed; and instead of potassium, I obtained only very limited quantity of a black substance, which choked up a part of the iron tube connected
with the iron bottle. This black substance was hastily transferred to a dry
bottle, which was then well corked. A small part of it was in powder, but the
greater part in little lumps slowly decomposed water, evolving only very
minute globules of gas; others decomposed that fluid very rapidly, producing all the gas they would furnish, with nearly the same facility as potassium would have done, under similar circumstances. The gas, thus slowly
produced, was on examination found to be hydrogen; whilst the gas rapidly
evolved, possessed properties so different from any other known gas, as to
entitle it to be regarded as a new combination.*

*The new gas was obtained by action of pure water on the black sub-
stance.*

When the new gas was mixed with about six times its volume of air, it exploded, when kindled, producing a white flame and a whistling sound. One measure of the new gas being mixed with about ten measures of air in a tube, and kindled, produced a loud explosion, accompanied by a blue flame, which pervaded nearly the whole length of the tube. One measure of the new gas, and nineteen of air, burned rapidly with a blue flame.

The new gas forms with oxygen a powerful explosive mixture, especially when the volume of the latter is about three or four times that of the former. In exploding a mixture of this kind in a detonating tube about half an inch in diameter, and nearly one-third of an inch thick, the tube was shattered in pieces by the violence of the shock, though the volume of new gas did not exceed 5/100 of a cubic inch.

When chlorine is brought in contact with the new gas, instant explosion takes place, accompanied by a large red flame, the deposition of much carbon, and condensation (to a certain extent) of the two gases; and these effects occur in the dark, and are of course quite independent of the action of the suns's rays, or of light.

Safety Precautions

Mixtures of acetylene and air are explosive. Safety glasses and protective gloves must be worn!

Apparatus

Tin can (volume about 1 liter) with a lid which is put on by pressing (not a screw top!) and a hole 0.5 cm in diameter in its side, long wooden splint, cigarette lighter, safety glasses, protective gloves.

Chemicals

Calcium carbide, water.

Experimental Procedure

About 5 pea-sized pieces of calcium carbide are placed in the can; one to two milliliters of water are added and the can closed by pressing the lid on hard. The can is stood vertically on the table with the lid uppermost. After waiting for about 15 seconds the wooden splint is ignited and the burning splint held next to the hole in the side of the can. The experimentalist must make sure that neither his head nor his other hand are in the vicinity of the can, as the lid is often hurled into the air with great force because of the violence of the explosion. A larger amount of calcium carbide should not be used; if the air in the can is almost completely displaced by acetylene there is either no explosion at all or only a very weak one.

Explanation

Acetylene reacts with the oxygen in the air to form CO, CO_2 and H_2O. Depending on the temperature and the oxygen content, the following reactions can occur:

$$C_2H_2 + 3/2\ O_2 \rightarrow 2\ CO + H_2O$$
$$C_2H_2 + 5/2\ O_2 \rightarrow 2\ CO_2 + H_2O$$

Waste Disposal

The residue is dissolved in a little dilute hydrochloric acid and poured down the drain.

109

Oxyhydrogen Gas in Soap Bubbles

Soap Bubbles

From years of study and of contemplation
An old man brews a work of clarity,
A gay and involuted dissertation,
Discoursing on sweet wisdom playfully.

An eager student bent on storming heights
Has delved in archives and in libraries,
But adds the touch of genius when he writes
A first book full of deepest subtleties.

A boy, with bowl and straw, sits and blows,
Filing with breath the bubbles from the bowl.
Each praises like a hymn, and each one glows;
Into the filmy beads he blows his soul.

Old man, student, boy, all these three
Out of the Maya-foam of the universe
Create illusions. None is better or worse.
But in each of them the Light of Eternity
Sees its reflection, and burns more joyfully.

Hermann Hesse, "The Glass Bead Game"

In 1863, *H. Sainte-Claire Deville* reported on the decomposition of water in "Annalen der Chemie und Pharmazie":

If one pours 1 to 2 kilograms of molten platinum into water, as Debray *and I have often done, one observes the plentiful release of an explosive gas consisting of hydrogen together with a certain amount of nitrogen, the latter having been dissolved in the water and liberated by the heat. This is a repeat on a large scale of an experiment carried out by* Grove, *in which water was decomposed by agitation with heated platinum at a temperature still well below its melting point; this experiment was the starting point for my investigations into decomposition.*

How does it transpire that platinum melts so readily through the influence of heat developed by the combination of hydrogen with oxygen; and that molten or merely white-hot platinum decomposes water?

This strongly heated vapor partially decomposes, and in proportion to the force of decomposition corresponding to the temperature of the melting platinum. The fraction subject to decomposition, and whose volume relative to its weight is significant, cools very rapidly, because it attempts to rise to the surface of the water, while the platinum rapidly sinks, and the rate of cooling is such that a portion of the detonating gas avoids recombination. This implies only that complete combustion of a limited amount of detonating gas requires a limited amount of time; and the fact that this is true is proven by the effect of metal gauze on burning gases.

But why does not all the detonating gas become water again in the course of cooling? This is due to two factors.

The first, a purely physical one, is also the cause of a very well-known fact: the incombustibility of detonating gas when the latter is admixed with a large quantity of an inert gas such as carbonic acid or nitrogen. Such a mixture indeed resists the effect of an electric spark, and fails to ignite upon contact with a burning object. But it cannot be passed slowly through a tube filled with red-hot porcelain fragments without those elements that are capable of binding becoming completely associated with each other.

Thus there is another factor, and this is a purely mechanical one; namely, the rate at which the gases in my apparatus pass through the porcelain tube; this is the basis for the rate of cooling or return to that temperature at which oxygen and hydrogen, diluted in a large amount of carbonic acid, no longer combine with each other.

Safety Precautions

Safety glasses must always be worn for this experiment!

Apparatus

Large test tube with connecting tube at the side, stopper with 2 nickel electrodes, two pieces of electric cable, stand, clamp and boss, transformer, gas offtake tube, small porcelain dish, wooden splint 30 cm in length, safety glasses, protective gloves.

Chemicals

1 mol/L H_2SO_4, soap solution.

Experimental Procedure

The test tube is filled with 1 mol/L H_2SO_4 up to the level of the side tube and closed at the top by means of a stopper through which the two nickel electrodes pass. 20 V DC are now applied. The oxyhydrogen gas thus generated is passed into the soap solution in the porcelain dish via a gas offtake tube which is attached to the side tube. After a while a ball of foam as big as a fist is formed. The DC current is turned off and the porcelain dish removed to a distance of at least a meter from the gas generating apparatus. The foam in the porcelain dish is ignited using the burning wooden splint. The oxyhydrogen gas in the foam reacts with a loud bang to give water. Sometimes the porcelain dish is broken by the force of the explosion.

It is important in this experiment to take great care that the gas generator (test tube with nickel electrodes) is removed to a sufficient distance in order to make sure that the oxyhydrogen gas in it does not also explode, as otherwise glass splitters and sulfuric acid will fly across the room! The experiment can be repeated several times.

Nitrogen Triiodide

In the observation of phenomena, the conduct of analyses, and other determinations, one is urgently warned against being influenced in any way by theories or other miscellaneous preconceived opinions.

Emil Fischer

As early as 1813 *B. Courtois* reported (*Ann. Chim.* **1813**, *88*, 309) that a black powder is formed when aqueous ammonia reacts with iodine. This so-called *iodure d'azote* (nitrogen iodide) is not a single homogeneous compound, but a mixture of nitrogen triiodide and addition compounds between ammonia and iodoamines. Other authors later obtained a compound of the composition $NI_3 \cdot NH_3$ using various methods.[1]

Safety Precautions

This experiment may only be carried out by experienced chemists. Nitrogen iodide is extremely explosive. Safety glasses and protective gloves must be worn!

Apparatus

Large test tube, stand, funnel containing a fluted filter paper, 500-mL Erlenmeyer flask, stick 2 meters in length, safety glasses, protective gloves.

Chemicals

Iodine, concentrated aqueous ammonia solution, ethanol, diethyl ether.

Experimental Procedure

A pinch of finely powdered iodine is placed in the test tube with 10 mL of the concentrated ammonia solution and either shaken well or allowed to stand overnight. If it is allowed to stand the test tube must be stoppered. The precipitate is then filtered off, excess iodine removed by washing with a few mL of ethanol and the residue treated with 5 mL of diethyl ether to dry it rapidly. The filter paper is removed from the funnel at once and carefully placed on a stable support.

After about an hour the powder on the filter paper is touched with the long stick; a violent explosion takes place and violet iodine vapor is formed.

Explanation

The following equation can be formulated for the decomposition of the nitrogen iodide:

$$2 \, NI_3 \cdot NH_3 \rightarrow 3 \, I_2 + N_2 + 2 \, NH_3$$

The high tendency for formation of the nitrogen molecule is responsible for the violence of the reaction.

Waste Disposal

Iodine residues should be stirred with a litte Na_2SO_3 solution and flushed down the drain.

Reference

1 Pure NI_3 can be prepared as described by I. Tornieporth-Oetting, T. Klapötke, *Angew. Chem.,* **1990,** *102,* 726; *Angew. Chem. Int. Ed. Engl.* **1990,** *29,* 677.

111

Minting Coins
using Potassium Chlorate

A chemistry teacher writes as follows about his experiences when experimenting with potassium chlorate:

Once at the beginning of my service as a chemistry teacher I mixed I thought very carefully potassium chlorate and phosphorus with the aid of a horn spoon. For three days my ears were ringing. The open dish was shattered, the shards had penetrated the desk, and despite a zealous search absolutely nothing was to be found of the horn spoon except the handel, which was in my hand.

Prior to the introduction of coins most of the ancient people used metal rings, bars or other forms for payment. Thus *Agatharchides* reports:

Through the middle of their country there flows a river that carries golden sand and apparently possesses such riches that the mud collected at its outlets glistens over a wise expanse. However, the inhabitants of the region know nothing about processing it (unlike their neighbors). They find much gold, which they did in the flat regions of the land not scientifically and artistically by smelting it from the sand, but rather as preexisting material, what the Greeks appropriately described as achieved without fire. The mallest piece is the size of a grain of wheat, a medium-sized one is like a medlar, and a large one like a walnut. They pierce the gold and wear it in alternation with transparent stones around the wrist and the neck, and trade it cheaply with their neighbors. Thus, they trade ore for three times its weight in gold, but iron for twice the weight, whereas silver counts for ten parts of gold.

C. Müller, "Geographi Graeci minores", Vol. I (1855)

The first coins were minted about 600 B.C. in the Greek cities of western Asia Minor; at that time these were subject to the Lydians, who are thus considered to be the discoverers of coins.

Safety Precautions

$KClO_3$ is a strong oxidising agent and reacts explosively with oxidisable substances such as sulfur, phosphorus or sugar. Safety glasses must be worn and a safety screen used when experimenting with $KClO_3$! CS_2 is highly toxic.

Apparatus

Flat iron block, aluminum sheet (0.1–0.3 mm thick), coin, 25-mL Erlenmeyer flask, dropping pipette, tweezers, filter paper, safety glasses, protective gloves.

Chemicals

$KClO_3$, CS_2, white phosphorus.

Experimental Procedure

The coin is placed on the iron block and the sheet aluminum laid over it. On the latter, exactly above the coin, is placed a piece of filter paper crumpled tightly to the size of a pea; this in turn is covered with 3 g of $KClO_3$. The $KClO_3$ is carefully impregnated with a solution of white phosphorus in carbon disulfide (0.1 g of P_4 in 2 mL of CS_2) using a dropping pipette. After about 5 to 10 minutes (depending on the temperature of the room) the evaporation of the carbon disulfide is complete and the phosphorus reacts explosively with the $KClO_3$. The force of the explosion causes the imprint of the coin to be stamped on the aluminum sheet. If the reaction does not start of its own accord, it can be initiated by applying a burning candle fixed to the end of a long wooden pointer.

Combustion with Emission of Sparks

Paracelsus gave a detailed description of fire:

> *Fire always requires a substrate; it cannot exist without air, moisture, and earthy vapors. The breath increases the flame, lamps consume more oil in a draft, and wood burns more rapidly. Coal and wood burn only in a draft, because they have narrow pores that the draft opens and makes more pervious to the fire. Flame is purer when it contains no aqueous or earthy components that lead to smoke and vapor.*
>
> *Based on F. Wimmer (1866)*

Wöhler also had problems with chemical experiments when he was at school, as is reported by *Kahlbaum (F. Wöhler, "Ein Jugendbildnis"* in letters to *Hermann von Meyer,* 1900):

> *It does not always occur without risk to house and home, and also to one self. Thus, the young laboratory technician prepares chlorine, but fails to secure the apparatus, and is nearly suffocated by the poisonous gases. At the same time he conducts an experiment with phosphorus in his room, and in the process burns his left hand so badly that the scars will still be apparent on his hand in old age. In the process, the phosphorus particles splatter everywhere in the room, and when he went to bed at night after extinguishing the light, he could everywhere see flaring lights; he became frightened and rushed to his father, who consoled and calmed the upset youth by explaining to him the lack of danger associated with the phenomenon.*
>
> *Operations requiring larger quantities of heat were carried out in the courtyard in the laundry, and an old graphite furnace donated by the master of the mint, Bunsen (a relative of the famous chemist), served as an oven; but one fine day in the course of one of the experiments this improvised fireproof laboratory also went up in flames.*

Safety Precautions

Safety glasses and protective gloves should be worn. The sparkler should be ignited at a distance of at least a meter from the reaction flask in order to ensure that no sparks can fly into the flask prior to the experiment. Care must be taken to make sure that neither hands nor face are directly above the flask when the sparkler is thrown in.

Apparatus

1-L beaker, 250-mL Erlenmeyer flask, dropping pipette, long-stem funnel, fire-resistant support 50 × 50 cm, sparkler, protective gloves, safety glasses.

Chemicals

Ignition powder for cartridges (can only be obtained from specialised dealers), gunpowder, magnesium, iron and aluminum powder.

Experimental Procedure

1 g of fine gunpowder, 8 g of (smokeless) ignition powder and 0.5 g each of magnesium, iron, and aluminum powder are placed in the 250-mL Erlenmeyer flask. The contents are now mixed well by shaking the flask and the latter placed in the beaker. The sparkler is ignited by means of a cigarette lighter and carefully dropped into the flask. The mixture immediately begins to burn; the sparks fly up to a height of 1.5 meters but go out during flight. The residue falls into the flask or to the floor within a few cm of the flask and causes no harm. This experiment can safely be carried out indoors. The Erlenmeyer flask and the beaker normally are not damaged (colored figure 32).

Explanation

Gunpowder contains sulfur, potassium nitrate and charcoal. In the mixture the gunpowder and the ignition serve as igniting charge and propellant for the metals respectively; the burn brightly in air to give their oxides MgO, Fe_2O_3 and Al_2O_3.

Waste Disposal

The metal oxides should be transferred to the container used for collecting less toxic inorganic waste.

The Charcoal Dance: Reactions of Charcoal and Sulfur with Fused Potassium Nitrate

Six things belong to a chemist:
Working day and night,
Stoking the fire continuously,
Suffering smoke and fumes,
Infecting oneself,
Losing sight and health,
And finally perceiving success in a troubled heart.

From an anonymous satirical poem

Safety Measures

Safetey glasses and protective gloves should be worn. The reaction is often so violent that the piece of charcoal is shot out of the test tube, so that the latter should not be pointed towards the audience. The charcoal burns harmlessly if the opening of the test tube is covered by a piece of copper gauze.

Apparatus

Large test tube, gas burner, stand, crucible tongs, porcelain dish, safety glasses, protective gloves.

Chemicals

Potassium nitrate, charcoal, piece of sulfur.

Experimental Procedure

5 g of potassium nitrate are fused in the test tube using a gas burner. A porcelain dish is placed underneath the test tube in case the latter should break. A pea-sized piece of charcoal is added when gas bubbles are visible in the fused salt mass. The charcoal burns brightly and does a "dance" on the surface of the fused salt.

The charcoal can be replaced by a piece of sulfur, which then burns with a bright flame to give sulfur dioxide. In this case the experiment must be carried out in a well-ventilated hood.

Explanation

At lower temperatures KNO_3 is reduced by the charcoal to KNO_2, CO and CO_2 being formed. The following somewhat more complicated reaction occurs at the temperature of the fused KNO_3.

$$3 \; C + 2 \; KNO_3 \rightarrow K_2CO_3 + N_2 + CO + CO_2$$

Potassium nitrate and elemental sulfur react to give sulfur dioxide.

Waste Disposal

The salts are dissolved in water and the solution poured down the drain.

114

Dancing Fire

Ring round ring I forthwith drew,
Wondrous flames collected there;
Herbs and bones in order fair,
Till the charm had work'd aright.

Johann Wolfgang von Goethe, "The Treasure-Digger" (1797)

According to *Empedocles* (490–430 B.C.) the universe consists of four elements: *fire, air, earth* and *water*. These elements are held together by two mystic forces. He believed love to be the force which held the elements together and hate to be that which caused them to separate. This concept was one of the cornerstones of chemistry for almost 2000 years.

Let us suppose that we could return to the time when fire was discovered. How did man live then? A good description is provided by *C. Sagan* in the book *"Our Cosmos"* (1989).

One day there was a storm, with much lightning and thunder and rain. The little ones are afraid of storms. And sometimes so am I. The secret of the storm is hidden. The thunder is deep and loud; the lightning is brief and bright. Maybe someone very powerful is very angry. It must be someone in the sky, I think.

After the storm there was a flickering and crackling in the forest nearby. We went to see. There was a bright, hot, leaping thing, yellow and red. We had never seen such a thing before. We now call it flame. It has a special smell. In a way it is alive. It eats food. It eats plants and tree limbs and even whole trees, if you let it. It is strong. But it is not very smart. If all the food is gone, it dies. It will not walk a spears throw from one tree to another if there is no food along the way. It cannot walk without eating. But where there is much food, it grows and makes many flame children.

One of us had a brave and fearful thought: to capture the flame, feed it a little, and make it our friend. We found some long branches of hard wood. The flame was eating them, but slowly. We could pick them up by the end that had no flame. If you run fast with a small flame, it dies. Their children are weak. We did not run. We walked, shouting good wishes. Do not die, we said to the flame. The other hunterfolk looked with wide eyes.

Ever after, we have carried it with us. We have a flame mother to feed the flame slowly so it does not die of hunger. Flame is a wonder, and useful too; surely a gift from powerful beings. Are they the same as the angry beings in the storm?

The flame keeps us warm on cold nights. It gives us light. It makes holes in the darkness when the Moon is new. We can fix spears at night for tomorrows hunt. And if we are not tired, even in the darkness we can see each other and talk. Also a good thing! Fire keeps animals away. We can be hurt at night. Sometimes we have been eaten, even by small animals, hyeanas and wolves. Now it is different. Now the flame keeps the animals back. We seen them baying softly in the dark, prowling, their eyes glowing in the light of the flame. They are frightened of the flame. But we are not frightened. The flame is ours. We take care of the flame. The flame takes care of us ...

After we found the flame, I was sitting near the campfire wondering about the stars. Slowly a thought came: The stars are flame, I thought. Then I had another thought: The stars are campfires that other hunterfolk light at night.

The stars give a smaller light than campfires. So the stars must be campfires very far away. But, they ask me, how can there be campfires in the sky? Why do the campfires and the hunter people around those flames not fall down at our feet? Why don't strange tribes drop from the sky?

Safety Measures

Safety glasses and protective gloves must be worn. Mixtures of methane and air are explosive: a fire extinguisher should be kept at hand.

Apparatus

Dewar vessel, stand, clamp and boss, two large test tubes, glass tube, rubber tubing, safety glasses, protective gloves.

Chemicals

Liquid nitrogen, methane from pressure cylinder.

Experimental Procedure[1]

10 mL of water are added to one of the test tubes; the level of the meniscus serves as the reference mark for the other test tube. This level is marked on the latter with a colored pencil. The marked test tube is fastened to the stand by means of the clamp and boss; its lower half is cooled by immersing it in the liquid nitrogen contained in the Dewar vessel. During the cooling process

methane is passed into the test tube from the pressure cylinder via the glass tube and rubber tubing. The glass tube should reach to the bottom of the test tube. Within about 5 minutes 10 mL of liquid methane condense out (until the mark has been reached); this is accompanied by a characteristic gurgling noise. The test tube must then be removed immediately from the liquid nitrogen to avoid condensation of oxygen from the air (the boiling point of methane is − 164 °C, that of oxygen − 183 °C). The glass tube and rubber tubing are also removed at once.

The gas in the test tube is then ignited by holding a cigarette lighter near its upper edge. The methane flame can be between 2 and 25 cm high, depending on the rate of evaporation. The room is darkened, the boss loosened and the burning methane poured slowly from a height of about 10 cm on to the demonstration table; the latter must have a raised edge, so that the methane cannot run down its sides. The experiment can also be carried out on a polished PVC floor if no persons or flammable objects are in the immediate vicinity. The behavior of the methane on reaching the floor can be simulated beforehand using the corresponding amount of liquid nitrogen. The burning drops of methane glide across the tabletop, because a thin layer of gas (which is a poor conductor of heat) forms between the liquid drop and the surface of the table. This is known as the *Leidenfrost* effect.

The audience is normally very impressed by the "dancing fire" in the darkened room.

Explanation

Methane burns in air according to the equation:

$$CH_4 + 2\ O_2 \rightarrow CO_2 + 2\ H_2O$$

Reference

1 D. M. Stamm, *J. Chem. Educ.*, **1992**, *69*, 762.

115

A Burning Gel

In that which we see, in that wich we find,
There may be both error and truth combined.

Sigmund Freud

Apparatus

400-mL beaker, two 250-mL beakers.

Chemicals

Calciumdiacetate, ethanol, 1 mol/L NaOH, alcoholic phenolphthalein solution.

Experimental Procedure

60 g $Ca(OOCCH_3)_2$ are dissolved in 200 mL of water in a 250-mL beaker; the solution must be saturated, and additional calcium diacetate can be added if necessary. Sodium hydroxide solution is added until the solution shows a weakly basic reaction with phenolphthalein. 40 mL of this solution are transferred to the larger beaker and treated with 300 mL of ethanol and 2 mL of the phenolphthalein solution. This mixture is now poured from one beaker to another until a pink gel is formed. The room is darkened and the gel ignited (see colored figure 33). If a metallic vessel is used, 200 mL of ethanol and 40 mL of the saturated calcium diacetate solution are poured into it together with the phenolphthalein solution; stirring is then not necessary, but after about 10 seconds the gel forms and can be ignited as described above.

Explanation

The addition of ethanol to the saturated calcium diacetate solution decreases the solubility of the salt, which quickly precipitates, forming a highly crosslinked lattice within the liquid and taking up the alcohol. A gel is thus formed, which burns as described above.

Waste Disposal

The solid residue is dissolved in water and poured down the drain.

Borates

The flame fighting the wind rises higher.

Michelangelo Buonarotti

Apparatus

1-L round-bottomed flask with a one-hole rubber bung, glass tube 1 m length passed through a rubber bung attached at one end, stand with clamps, tripod, wire net, gas burner, boiling chips, measuring cylinder, wide-necked test tubes, spatula, dropping pipette, safety glasses, protective gloves.

Chemicals

Boric acid, borax $Na_2B_4O_7 \cdot 10\ H_2O$, methanol, ethanol, concentrated sulfuric acid.

Experimental Procedure

30 g of boric acid are introduced into the round-bottomed flask, followed by 150 mL of methanol; a few boiling chips and 2–3 mL of H_2SO_4 are added. The round-bottomed flask is now fixed to the stand and placed on the wire net on the tripod; the rubber bung bearing the glass tubing is inserted. The latter should end just below the bung, and should be fixed to the stand by means of a second clamp. The mixture is heated until it boils and the vapor which escapes from the glass tube is ignited. A beautiful green flame is observed, the height of which can be adjusted by varying the rate of heating. The process can be controlled so that the flame is formed 30–40 cm above the top of the glass tube and reaches a height of up to 2 meters. If the experiment is repeated after inserting a little glass wool at about the middle of the glass tube no flame is observed at first.

   The green color of the flame formed when methyl borate burns can be demonstrated in a simpler form by carrying out the experiment in a wide-necked test tube (see colored figure 34); however, the flame does not reach such a great height. If boric acid is replaced by borax, the crystalline hydrate of disodium tetraborate, the same effect is obtained when a little more sulfuric acid is added. However, if methanol is replaced by ethanol a green flame is formed only if boric acid is used; in this case the use of borax leads to a yellow coloration which takes on a somewhat green hue only if very large amounts of sulfuric acid are used.

Explanation

Boric acid reacts with methanol in the presence of H_2SO_4 as a catalyst to form the corresponding methyl ester (eqn 1):

$$(1)\ B(OH)_3 + 3\ CH_3OH \rightarrow B(OCH_3)_3 + 3\ H_2O$$

This ester gives the green color. The glass wool adsorbs the ester vapor, so that no color is observed when it is present. Sulfuric acid sets boric acid or its anhydride free from borax, thus permitting the formation of the ester. Ethanol does not react so well.

Waste Disposal

The residue is transferred to the container used for collecting flammable organic solvents.

Ethyl Acetate

You may object that by speaking of simplicity and beauty I am introducing aesthetic criteria of truth, and I frankly admit that I am strongly attracted by the simplicity and beauty of the mathematical schemes which nature presents us. You must have felt this too: the almost frightening simplicity and wholeness of the relationship, which nature suddenly spreads out before us ...

Werner Heisenberg (1901–1976), "Physics and Beyond"

Apparatus

500-mL round-bottomed flask with reflux condenser, stand with clamps, hotplate, glass jar, 250-mL Erlenmeyer flask, 50-mL measuring cyclinder, dropping pipette, wide-necked test tube, boiling chips, pipette, safety glasses, protective gloves.

Chemicals

Acetic acid, ethanol, concentrated sulfuric acid, 0.5 mol/L NaOH, 1 % phenolphthalein solution, strips of polystyrene foam.

Experimental Procedure

100 mL each of acetic acid and ethanol are mixed in the round-bottomed flask. 20 mL of concentrated sulfuric acid are added, together with a few boiling chips, and the mixture heated at reflux for about 10 minutes (see figure). After cooling the mixture is poured into a glass jar; two samples are taken from the upper ester phase and placed in the Erlenmeyer flask and the test tube. The fruity smell of ethyl acetate can soon be noticed throughout the room. If a strip of polystyrene foam is dipped into the test tube containing the ester the polymer rapidly dissolves. The sample (10 mL) in the Erlenmeyer flask is treated with 40 mL of water and 4 mL of the 0.5 mol/L NaOH solution; a few drops of phenolphthalein solution are added. The mixture is heated on the hotplate. After a while the ester smell disappears, the red color of the indicator becoming lighter and finally disappearing: the ester has been hydrolysed.

Explanation

One of the relevant examples in chemistry which can be treated in terms of the application of the rules of mathematics is the Law of Mass Action. This simple and beautiful mathematical treatment of equilibrium processes is of great scientific and economic importance (e.g. for the calculation of yields!). The synthesis of esters is a typical reversible process and occurs as shown in eqn. 1:

$$(1)\ CH_3COOH + C_2H_5OH \rightleftharpoons CH_3COOC_2H_5 + H_2O$$

We shall not refer here to the application of suitable mathematical examples to this equilibrium process, as these are dealt with in detail in textbooks of physical chemistry. The ester, which is present in the upper phase of the heterogeneous system formed, is hydrophobic and readily dissolves plastics, fats, resins and other high molecular mass organic substances. The process in which it is formed is reversed when it is treated with sodium hydroxide.

The acetic acid in the homogeneous system neutralises the amount of sodium hydroxide added, so that the phenolphthalein is decolorised. A solid residue of sodium acetate is obtained when the solution is evaporated to dryness. The ester formation can be promoted by the addition of small amounts of concentrated sulfuric acid (or P_4O_{10}) as these bind the water which is formed (eqn. 1)

Waste Disposal

The remainder of the reaction solutions can be poured down the drain, while the contents of the test tube should be transferred to the container used for collecting halogen-free organic solvents.

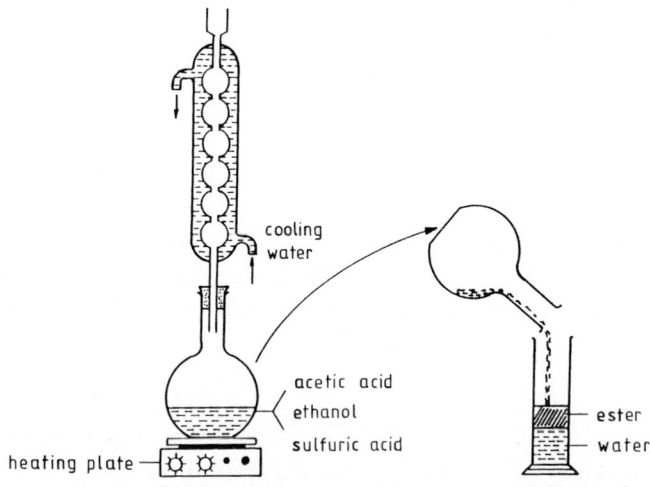

Experimental setup for the synthesis of ethyl acetate

Esters as Natural Perfumes

The English translation of *Titus Lucretius Caros*, Poem *"De Rerum Natura"* *(The Way Things Are)* is as follows:

Now then I have shown that things can never be created
From nothing, and that no created thing
Can ever be called back to nothingness.
You may, perhaps, begin to doubt my lessons
Since atoms are too small to see, but listen,
You must admit that there are other bodies
Existing but invisible.
...
I tell you again and again, the winds are bodies
Invisible, they are rivals of great rivers
In what they do and are, though rivers, of course,
Are something we can see.

And what of odors?
We sense them, but we never see them coming
Toward our nostrils; we do not look at heat,
Apprehend cold with our eyes, we are not accustomed
To witness voices, yet all these things, by nature,
Must be material, since they strike our senses.
Nothing can touch or be touched, excepting matter.

Esters derived from small- to medium-chain carboxylic acids and small- to medium-chain alcohols are the so-called fruit ethers. These are present in fruits and ethereal oils and are characterised by their pleasant aromas. A large variety of synthetic aromatic principles for use in the perfume industry can be synthesised from suitable carboxylic acids and alcohols.

Apparatus

Four wide-necked test tubes, test tube holder, beaker to contain the test tubes, gas burner, tripod with wire gauze net, dropping pipettes, safety glasses, protective gloves.

Chemicals

Concentrated sulfuric acid, methanol, ethanol, 1-pentanol, 2-pentanol, acetic acid, isovaleric acid (3-methylbutanoic acid), benzoic acid, salicylic acid, anthranilic acid.

Experimental Procedure

(I) Amyl Acetate

4 mL of 1-pentanol and 4 mL of acetic acid are mixed in a test tube and treated carefully with 1–2 mL of concentrated sulfuric acid. The mixture is warmed for a short time in the flame of the gas burner, the ester (b.p. 142 °C) which is formed occurs in nature and smells of pears.

(II) Isoamyl Valerate

It is advisable to carry out this experiment in a well-ventilated hood because of the unpleasant smell of the carboxylic acid (valerian!).

A mixture of 3 mL of 2-pentanol, 3 mL of isovaleric acid and 1 mL of concentrated H_2SO_4 in a test tube is heated in the flame of the gas burner; after a while the obnoxious smell of the starting materials is replaced by a pleasant banana-like smell. The ester which is formed (b.p. 190 °C) is present in bananas and is used in artificial pineapple and peach oils.

(III) Methyl Benzoate

1 g of benzoic acid is placed in a test tube: 4 mL of methanol and 1.5 mL of concentrated sulfuric acid are added and the mixture subjected to mild heating. After a while a colorless, pleasant-smelling substance (b.p. 198 °C) is formed; its non-chemical name is niobe oil. Ethyl benzoate, which smells of cloves, can be formed in an analogous manner or by treating benzoyl chloride with ethanol.

(IV) Methyl Salicylate

2 g of salicylic acid and 4 mL of methanol are heated for 6–7 minutes in a water bath (beaker containing boiling water) with 4 drops of concentrated sulfuric acid. After a short time the smell of oil of wintergreen becomes noticeable; this oil, which contains 96–97 % methyl salicylate, is manufactured on an industrial

benzoic acid ester

(arom of cloves)   (niobe oil)

Benzoic acid esters as perfumes

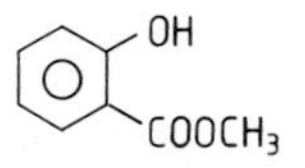

salicylic acid ester

(oil of wintergreen)

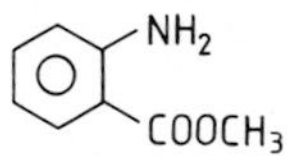

anthranilic acid ester

(arom of orange blossoms)

Methyl esters of benzoic acid derivatives

scale and used in the treatment of rheumatism. Methyl anthranilate, which smells of orange blossoms and melts at 24 °C, is obtained in a similar manner (see formulae).

Explanation

The esters are formed in the reaction between the relevant carboxylic acids and alcohols, one of the components being used in excess.

The concentrated sulfuric acid catalyses the reaction and binds the water formed. The reaction rate is increased by raising the temperature. However, the reaction mixture must be heated at reflux for several hours if large amounts of the ester are required. The reaction products are purified by distillation and characterised by means of modern analytical procedures.[1] The esters are generally extracted from natural products by a careful steam distillation, which is followed by separation of the components of the mixture.

Waste Disposal

The reaction products are hydrolysed by adding sodium hydroxide solution, diluted and poured down the drain.

Reference

1 A. Mosandl, *Kontakte*, **1992**, *3*, 38.

Reactive Aldehydes

In the year 1835 *Liebig* isolated acetaldehyde, which had previously been obtained in an impure form by *Scheele* (1782) and *Döbereiner* (1821). The name *"aldehyde"* comes from *"al*cohol *dehyd*rogenatus*"* is due to *Liebig* and was suggested by *Poggendorf*. Formaldehyde and the other aliphatic aldehydes have chemical properties similar to those of acetaldehyde, the most important being a marked ability to reduce a large number of substances and to undergo addition reactions with other unsaturated compounds. The aromatic aldehydes, which are derived from benzaldehyde (a liquid which can be isolated from bitter almonds) resemble the aliphatic compounds in their reactivity. However, their tendency to form more stable oligomeric units is not so strong. The aromatic rings determine the properties of this series of aldehydes to a great extent, as is shown by their persistent and quite pleasant aromas. They are thus often used as artificial fragrances. The factor common to all aldehydes is their *functional group-CHO*, the analytical identification of which we shall demonstrate in the following experiments.

Apparatus

Three 250-mL Erlenmeyer flasks, 2-L beaker, 100-mL round-bottomed flask with fractionating column, Liebig condenser and distillate receiver, water bath, dish with cooling mixture, widenecked test tubes with stand, suction flask with small glass frit, stand with clamps, gas burner, measuring cyclinder, glass rods, dropping pipettes, spatula, micro-scale melting point apparatus, safety glasses, protective gloves.

Chemicals

Benzaldehyde, salicylaldehyde, semicarbazide hydrochloride, paraldehyde, 1 % solutions of Na_2CO_3, NaOH, $AgNO_3$, NH_3, 4 % solutions of $Bi(NO_3)_3$ and Seignette salt $KNaC_4H_4O_6 \cdot 4\,H_2O$, concentrated H_2SO_4, ethanol, formaldehyde solution (highly diluted), $NaHSO_3$, parafuchsine.

Experimental Procedure

(I) Reductive Properties of Aldehydes

(1) 40 mL of the silver nitrate solution is treated with the same amount of sodium hydroxide solution in an Erlenmeyer flask; ammonia is added until the precipitate of $(AgOH)_{aq}$ which is initially formed redissolves completely. Samples are now placed in 4 test tubes as follows: 4 mL of formaldehyde solution (test tube 1), a mixture of 2 mL of benzaldehyde and 5 mL of ethanol (test tube 2), 3 mL of pure paraldehyde (test tube 3) and a mixture of 3 mL of freshly distilled acetaldehyde* with 3 mL of water (test tube 4). 5 mL of the $[Ag(NH_3)_2]^+$ solution are added to each test tube in the cold. The solutions in test tubes 1, 2 and 4 are colored black by the silver precipitate, while paraldehyde shows no reaction. A beautiful silver mirror is formed on the lower part of test tube 3 if it is warmed slightly.

(2) 20 mL of the 4 % solution of bismuth nitrate are treated in the second Erlenmeyer flask with 40 mL of sodium hydroxide solution and 20 mL of the Seignette salt solution until the resulting solution is completely homogeneous; if necessary more Seignette salt solution can be added. As described above, 4 samples of the aldehyde solution are treated with 5 mL of the bismuth salt solution and heated to boiling. After a while black, finely-divided bismuth is precipitated in test tubes 1, 2 and 4, while no reaction occurs in test tube 3.

(II) Reaction with Fuchsinesulfurous Acid

About 1 g of the dyestuff parafuchsine hydrochloride is dissolved in 1L of hot water in the large beaker. After the solution has cooled, sodium hydrogensulfite is added until the solution is colorless. The aldehyde test is carried out in 4 test tubes: 2 mL of the fuchsinesulfurous acid (to which a few crystals of $NaHSO_3$ have been added) are added dropwise to 2 mL of each of the samples (see above) in aqueous-alcoholic solution. After a few minutes the contents of test tubes 1, 2 and 4 are reddish violet in color, while those of test tube 3 remain colorless.

(III) Detection of the Aldehyde as its Semicarbazone

10 g of semicarbazide hydrochloride are stirred into 150 mL of water in the third Erlenmeyer flask. 3 mL each of benzaldehyde and salicylaldehyde are dissolved in 5 mL of ethanol in two test tubes and stirred with 5 mL of the semicarbazide solution after adding a few drops of 10 % soda solution. In each case

---

* Acetaldehyde is obtained as follows:
  30 g of paraldehyde are mixed with 2 ml of concentrated sulfuric acid. A few boiling chips are added and the mixture distilled from a water bath. The acetaldehyde, which boils at 21 °C, condenses in the well-cooled receiver. It can be used for the experiment either in a pure state or mixed with water. Paraldehyde is again formed if the acetaldehyde is allowed to stand for a long period of time.

a milky white or whitish yellow precipitate is formed immediately; this separates completely from the solution on further stirring. The mixtures are filtered through the glass frit by applying a vacuum and the residues recrystallised from a small amount of ethanol. White needle-shaped crystals are formed on cooling; their formation can be promoted by adding a few drops of ice cold diethyl ether. The crystals melt at 216 °C (benzaldehyde semicarbazone), and 229 °C (salicylaldehyde semicarbazone).

Explanation

In the course of the reactions described in section (I) the complex silver and bismuth ions are reduced to the metal by the -CHO group of the aldehyde. No reaction occurs if this group is not present (paraldehyde). In reaction (II) the SO_3 group of the fuchsinesulfurous acid is split off and adds to the aldehyde to form an α-hydroxysulfonic acid. The color of the fuchsine itself is now visible. All the "normal" reactions of an aldehyde are observed if acetaldehyde is set free from its trimer paraldehyde. Sulfuric acid catalyses both the forward and back reactions in this equilibrium (eqn. 1):

$$(1)\ (CH_3CHO)_3 \rightleftarrows 3\ CH_3CHO$$

Removal of the acetaldehyde by distillation shifts the equilibrium to the right

The formation of semicarbazones (III) is a reaction common to all aldehydes. For preparative reasons we have only described reactions involving aromatic aldehydes, since these afford semicarbazones which crystallise very readily (eqn. 2):

$$(2)\ R\text{-}CHO + H_2N\text{-}NH\text{-}CO\text{-}NH_2 \rightarrow R\text{-}CH{=}N\text{-}NH\text{-}CO\text{-}NH_2 + H_2O$$

The aldehyde can be identified very readily simply by determining the melting point of its semicarbazone. Aromatic aldehydes are in general pleasant-smelling substances. Thus benzaldehyde smells of bitter almonds, cinnamaldehyde of cinnamon, anisaldehyde of aniseed and vanillin of vanilla (see figure); the latter aldehyde can be obtained on an industrial scale by heating lignin with nitrobenzene and NaOH. Salicylaldehyde smells strongly of spiraea.

4 - methoxy benzaldehyde   3 - methoxy - 4 - hydroxy benzaldehyde

(anisaldehyde)   (vanillin)

Aldehydes as aromatic principles

Waste Disposal

The reaction mixtures from part (I) are oxidised with half-concentrated nitric acid; silver can be regenerated from the silver nitrate solution by adding zinc. Danger! Do not store ammoniacal silver nitrate solutions! The bismuth salt solutions are treated with soda. The precipitates are transferred to the container used for collecting less toxic inorganic waste, and the clear solution is poured down the drain, as are the solutions of the reaction mixtures from parts (II) and (III).

120

How to Dissolve Polystyrene Foam

*Once when he related that he had been well acquainted with Pontius Pilate
in Jerusalem, he described in minute detail the house of the governor, and
recited the dishes that had been served when he dined with him one evening. Cardinal de Rohan, who believed he was hearing tales of fantasy,
turned to the valet of the Count of Saint-Germain, an old man with white
hair and an honest countenance: "My friend" he said, "I find it difficult to
believe what your master is asserting. I dont mind if he is a ventriloquist;
and that he makes gold, fine; but that he is two thousand years old and was
supposedly acquainted with Pontius Pilate, that is too much. Were you
there?" "Oh, no, Monsignore", the servant replied inginously, "I have only
been in the service of my lord the Count for four hundred years".*

Collin de Plancy, "Dictionnaire infernal" Paris (1844)

Apparatus

Magnetic stirrer with stirring rod, 2-L beaker, safety glasses, protective gloves.

Chemicals

Acetone, polystyrene foam (as chips etc.).

Experimental Procedure

About 400 mL of acetone are placed in the beaker, which is standing on the
magnetic stirrer. The solvent is stirred rapidly and the polystyrene foam added.
The contents of a small sack will dissolve very rapidly (colored figure 35).

Waste Disposal

The solution is transferred to the container used for collecting halogen-free
organic solvents.

121

Hoarfrost in a Glass

Anyone who pays a little attention to the growth of plants will readily observe that certain of their external members are sometimes transformed, so that they assume – either wholly or in some lesser degree – the form of the members nearest in the series.

Thus, for example, the usual process by which a single flower becomes double, is that, instead of filaments and anthers, petals are developed; these either show a complete resemblance in form and color to the other leaves of the corolla, or they still carry some visible traces of their origin.

If we note that it is in this way possible for the plant to take a step backwards and thus to reverse the order of growth, we shall obtain so much the more insight into Nature's regular procedure; and we shall make the acquaintance of the laws of transmutation, according to which she produces one part from another, and sets before us the most varied forms through modification of a single organ.

Johann Wolfgang von Goethe,
"The Metamorphosis of Plants" (1790)

Material

Container made of fire-resistant glass, such as a large Erlenmeyer flask, spatula, glass rod, gas burner watchglass crystallising dish, plant, safety glasses, protective gloves.

Chemicals

Benzoic acid, diethyl ether.

Experimental Procedure

A small plant is fixed in position with a wire in such a way that it just remains standing. It is then placed in the glass flask, the bottom of which is covered by 10 g of benzoic acid. A suitable watchglass is placed over the flask (a small porcelain dish can also be used) and the Erlenmeyer flask with its contents heated on the wire net with a small flame. After a short time white "hoar frost" is seen to be formed on the plant.

Explanation

Benzoic acid, known since the 17th century as a sublimation product of gum benzoin, sublimes at 100 °C and is deposited on the surface of the plant as a molecular vapor; slight cooling thus leads to the formation of the white "hoar frost crystals".

Waste Disposal

Diethyl ether is added to the flask, which is closed with a stopper and shaken. The wire framework and the plant are removed, while the ethereal solution is poured into a crystallisation dish. The solution is allowed to evaporate in a well-ventilated hood, and the benzoic acid can be re-used. **Make sure no flames are present in the vicinity of the ethereal solution!**

122

Sulfur Crystals

It is not possible that any nature should be inferiour unto art, since that all arts imitate nature. If this be so; that the most perfect and generall nature of all natures should in her operation come short of the skill of arts, is most improbable. Now common is it to all arts, to make that which is worse for the betters sake. Much more then doth the common Nature do the same. Hence is the first ground of Justice. From Justice all other vertues have their existence. For Justice cannot be preserved, if either we settle our mindes and affections upon worldly things; or be apt to be deceived, or rash, and inconstant.

Marcus Aurelius Antoninus, "His Meditations" 11, 9

Heumann describes the *"crystallization of sulfurs"* in the following way:

Chunks of sulfur are melted in a porcelain crucible about 8 cm high and 6 cm wide which is heated over a gas lamp, and more are added until the crucible is about 2/3 full of molten sulfur. Should it happen to ignite, the crucible is covered with a lid for a few moments. The crucible is then removed from the fire and placed on an unheated sand bath, but not on the cold benchtop. It is left to stand undisturbed without a lid until sulfur on the surface is completely solidified and the outer wall of the crucible has cooled to such an extent that a piece of sulfur pressed against it does not become the least bit soft; the solid sulfur surface is then pierced with a hot rod, or an opening is cut in it with a knife, allowing as much as possible of the remaining liquid sulfur to flow out, when the crucible is inverted, into a porcelain dish filled with water. The sulfur crust is removed completely with the help of a knife, whereupon the inner walls of the crucible will be found to be covered with honey-yellow, transparent, prismatic crystals. After several days often within only a few hours these crystals become opaque and more white-yellow, in that they are transformed into the rhombic modification, even though retaining their prismatic external form.[1]

Sulfur has been known since the 16th century B.C., as is shown by the numerous references to be found in the *Old Testament*, the *Odysee* or *Pliny the*

Elder's most famous work *"Naturalis historia"*. The repetition of the above experiment, which describes the "crystallisation of sulfur" very clearly and in great detail, is highly recommended; it also provides stimuli for further study of the structure of sulfur.

Safety Precautions

All manipulations involving CS_2 and aromatic compounds must be carried out in a well-ventilated fume hood. CS_2 is extremely toxic and highly flammable.

Apparatus

Three 100-mL beakers, wide-necked test tubes and test tube holder, glass rods, heating plate, large petri dish, melting point apparatus, thermometer, balance, cooling bath, spatula, funnel with fluted filter paper, safety glasses, protective gloves.

Chemicals

Flowers of sulfur, CS_2, toluene, piperidine, distilled water, *p*-xylene.

Experimental Procedure

(I): 20 g of flowers of sulfur are suspended in 50 g of CS_2 in a beaker, and the mixture heated to its boiling point. It is then poured rapidly through a fluted filter paper into a well-cooled test tube. Yellow crystals of orthorhombic sulfur are formed at once (colored figure 36). The residue in the filter paper can be used for further experiments.

3 g of flowers of sulfur are suspended in 60 mL of toluene in another beaker and stirred for a few minutes just below the boiling point of the solvent. The hot solution is filtered rapidly through a fluted filter paper into a test tube, at the bottom of which pale yellow crystals slowly grow; these consist mainly of the monoclinic modification of sulfur (colored figure 37). The monoclinic modification is also predominantly formed during recrystallization from *p*-xylene at 120 °C (do not heat to boiling!).

Bright yellow sulfur crystals are formed when 3 g of the sulfur residues from the previous experiments are heated to boiling with 30 mL of *p*-xylene and the filtrate subjected to shock cooling in the cooling bath.

At room temperature the latter two modifications are transformed to orthorhombic sulfur, the thermodynamically most stable modification.

Blood-red homogeneous solutions are formed when any modification of sulfur is dissolved in piperidine.

The transformation of the orthorhombic modification to the monoclinic modification can be readily observed under the microscope when a micro melting point apparatus is used and the sulfur heated in the range 95–119 °C (the temperature at which melting commences). The reconversion of the monoclinic to the orthorhombic modification occurs very rapidly on cooling when 2–3 seed crystals are added.

(II): When a few grams of flowers of sulfur are heated in a test tube to above their melting point, the yellow color of the liquid becomes deeper up to about 160 °C, at which temperature the liquid is still mobile. Above this temperature the liquid turns brown, and at about 190 °C there is a sudden increase in its viscosity. If the heating is stopped for a moment, the dark brown sulfur can be seen to stick to the glass. The sulfur becomes less viscous when the temperature is increased further, and at 400 °C (just below its boiling point) it is no longer viscous. The liquid sulfur is now poured in a thin stream into a Petri dish full of cold water; it is transformed to a rubber-like brown-yellow modification which forms a fibre on pulling carefully with the help of a glass rod. After some time this viscous elastic material is transformed to the orthorhombic modification of sulfur.

Explanation

The yellow orthorhombic modification of sulfur is the thermodynamically most stable one. The crown-like structure of α-cyclo-S_8 was discovered in 1935. β-*Cyclo*-S_8 is monoclinic and is formed slowly above 95 °C, the transition temperature of orthorhombic sulfur. The third crystalline modification is γ-*cyclo*-S_8, which melts at 106.8 °C on rapid heating and can be prepared as discussed above. There is a large number of well-studied cyclic oligomers of sulfur; here we refer the reader to a standard textbook.[2] When sulfur heated to about 400 °C is subjected to quenching the product is polycatenasulfur, which exhibits rubber-like properties.

Waste Disposal

Any remaining sulfur can be reused. The aromatic hydrocarbons and the CS_2 can also be reused after purification by distillation.

References

1 Cited by O. Krätz, *Historisch-chemische Versuche*, Aulis-Verlag, Köln, **1987**, 233.
2 N. N. Greenwood, A. Earnshaw, *Chemistry of the Elements*, Pergamon Press, Oxford, New York, Toronto, Sydney, Paris, Frankfurt **1984**, 769.

Giant Crystals

Twelve gemstones are the symbols of perfection, holiness and belonging to the chosen, but there is also the *"secret word in the glow of bright stars" (Theodor Körner)*. In his poem *"Die Monatssteine"* he linked various beautiful minerals with the months of the year, as shown in the table below:[1]

	Gemstones of the month and their colors after *Theodor Körner* (1810)
January	*Hyacinth (red-brown)*
February	*Amethyst (violet)*
March	*Heliotrope (dark green)*
April	*Sapphire (blue)*
May	*Emerald (green)*
June	*Chalcedony (white /blue)*
July	*Carnelian (dark red)*
August	*Onyx (black/white)*
September	*Chrysolite (green)*
October	*Aquamarine (blue)*
November	*Topaz (yellow)*
December	*Chrysoprase (green)*

Apparatus

Two 300-mL beakers, 1-L beaker, 2-L beaker, watchglasses, funnels with fluted filter papers, stand with fixing device, glass rods, gas burner, tripod with wire gauze net, cross made of brass or steel with a roughened surface, balance, porcelain spoon, strand of wool, safety glasses, protective gloves.

Chemicals

$CuSO_4 \cdot 5\ H_2O$, $K_3[Fe(CN)_6]$, $KCr(SO_4)_2 \cdot 12\ H_2O$, $KAl(SO_4)_2 \cdot 12\ H_2O$, $FeSO_4 \cdot 7\ H_2O$, concentrated H_2SO_4, $(NH_4)_2SO_4$, distilled water.

Experimental Procedure

(I): 75 g of $CuSO_4 \cdot 5\ H_2O$ are suspended in 100 mL of distilled water in a 300 mL beaker and heated to boiling with constant stirring. The solution is filtered rapidly into the second 300 mL beaker, which is covered with a watch-glass and allowed to cool. After a while (sometimes at once) the crystallization process begins. It can be influenced by introducing a strand of wool, in which case a chain of crystals is formed along the wool. If no crystallization occurs after several hours, even with the help of the wool, a few seed crystals of copper vitriol are added; these cause immediate crystallization.[1]

(II): The metal cross is fastened upright in a 1-L beaker. A saturated aqueous solution of red prussiate of potash $K_3[Fe(CN)_6]$ is poured in until the beaker is almost full and the cross completely covered. The solution is prepared by heating a suspension of 500 g of the salt in 1 L of water; it is filtered while hot and poured into the beaker. During a few hours red crystals separate from the yellowish solution and cover the cross (see figure).

(III): The formation of alum crystals such as chrome alum is carried out as described in (I) above. 40 g of chrome alum or 30 g of $KAl(SO_4)_2 \cdot 12\ H_2O$ suffice when 100 mL of water are used. The octahedral crystals (see figure) are formed within several hours. If the isomorphous iron (III) alum is required, 20 g of $FeSO_4 \cdot 7\ H_2O$ are stirred with 30 mL of half-concentrated nitric acid; the solution is kept at boiling point for 15 minutes and a hot solution of 30 g $(NH_4)_2SO_4$ in 30 mL of water is added. The resulting solution is stirred for a

Cross covered with crystals of potassium ferricyanide

A particularly large crystal of chrome alum

further few minutes at the boiling point of water, and the precipitate remaining is filtered off and the solution left to cool. After a few hours octahedral crystals of iron (III) alum are formed; these should be almost colorless.

Explanation

All the salts which are used in this experiment dissolve better in warm water than in cold. Copper sulfate shows this most clearly. Crystal formation commences after a time when the hot saturated solutions are allowed to cool.

Waste Disposal

The crystals can be stored. The remaining solutions are treated with milk of lime. The residues are transferred to the container used for collecting less toxic inorganic substances.

Reference

1 F. Cherrier, *Chemie macht Spaß*, Part II, p. 11, Verlag J. F. Schreiber, Esslingen; Österr. Bundesverlag, Wien; Schwager and Steinlein, Nürnberg.

124

A Handkerchief Flambé, and How (Not) to Burn a Banknote

Science can never be related.
One can relate history,
But only write science.

Georg Christoph Lichtenberg

A dough prepared from 250 g of flour, 100 g of sugar, 3 eggs, 200 mL of boiled milk, 15 mL of cream, and a little salt is allowed to stand for 2 hours. Bake ca. 25 crêpes with 30 mL of oil and 30 g of butter.

Stir until frothy 250 g of butter and 150 g of sugar, to which is added the peel of 1/2 of a lemon and 1/2 of an orange. Flame it with 100 mL of Curaçao and 100 mL of cognac.

The sifted flour is mixed with the sugar and salt. The eggs are stirred in, followed by the previously boiled (but no longer hot) milk and the cream, after which the dough is left to stand at least 2 hours at room temperature.

The crêpes should be baked shortly before they are to be eaten: A little butter and oil are melted in a 12 cm pan at medium heat; somewhat more than 1 tablespoon of dough is placed in the pan and rapidly distributed by shaking, and the crêpe is turned as soon as its upper surface becomes firm. The best way to turn it is to toss it into the air (easily mastered with a little practice), or else one uses a metal spatula. The baking time is shorter for the second side. The crêpes should then be layered one above the other on a plate.

At the table, melt in a large pan about 1/6 of the butter, quickly add 4 crêpes one after another, turn them, fold them into half-circles, pour 1/6 of the previously prepared alcohol mixture over them, and flame them.

Heinz Maier-Leibnitz, Kochbuch für Füchse,
Piper Verlag Munich (1980)

Men write a great deal about the essence of material. I wish that material would begin to write about the human soul.

Georg Christoph Lichtenberg

*Bounteous is fire in its might
When man its power can guide aright,
And every art to mortals known
This Heaven-born agent's aid must own.
But fearful when from heaven sent
With loosened rein the element;
On headlong errand starting wild,
Nature's free and wayward child.*

*Friedrich Schiller (1759–1805),
"The Song of the Bell"*

Apparatus

Handkerchief, stand, clamp, iron rod, beaker, banknote, crucible tongs, safety glasses, protective gloves.

Chemicals

Ethanol, water, NaCl.

Experimental Procedure

A mixture of 50 mL of ethanol, small amount of NaCl (0.01 g) and 50 mL of water is prepared. The handkerchief is immersed in the alcohol-water mixture for a minute or so and fastened to a horizontal metal rod attached to the stand. The handkerchief is set on fire; after about 20 seconds it can be removed and is seen to be intact.[1]

The banknote is also immersed in this mixture; while it is still very wet it is held at one corner using the crucible tongs and lighted at its lower edge. It also "burns" without being damaged.

Explanation

Neither the handkerchief nor the banknote have suffered any damage, because only the alcohol burns. The flame has the characteristic yellow color of sodium. The boiling point of ethanol is 78 °C, while its flash point is 12 °C.

Waste Disposal

Larger amounts of ethanol should be transferred to the container used for collecting halogen-free organic solvents. Small amounts can be diluted with water and poured down the drain, as ethanol is relatively readily decomposed by microorganisms.

Reference

1 P. S. Bailey, C. A. Bailey, J. Andersen, P. G. Koski, C. Rechsteiner, *J. Chem. Educ.*, **1975**, *52*, 524.

A work such as this is actually never complete.

One must declare it to be complete when one has done all that is possible given the time and the circumstances.

Johann Wolfgang von Goethe, "Italian Journey" (1787)

Cited Personalities

Abu Mussah Dschafar al Sofi
  (Geber) 18
Adams, B. M. 144
Adams, D. 54
Agatharchides 294
Al-Biruni 139
Amenophis IV. 275
Aristotle 20
Aurelius, M. A. 137, 318
Avicenna 139

Bacon, F. 133
Bayen 83
Beckmann, E. 75
Bednorz, J. G. 240
Bell, A. G. 144
Belousow, B. P. 258
Belshazzar 112
Bergmann, T. O. 108
Berthollet, C. L. 120
Berzelius, J. J. 62, 85
Biringuccio, V. 48
Boerhaave, H. 117
Boisbaudran, P. E. Lecoq de 224
Boyle, R. 203
Brand, H. 42, 188, 203
Brandt, G. 108
Brassai, G. H. 249
Brechtel, F. J. 201
Bukatsch, F. 184
Bunsen, R. W. 243, 280
Busch, W. 33, 194

Caro, H. 76
Casriaroli, V. 219

Cavallo, V. C. 266
Cavendish, H. 266, 282
Chamisso, A. v. 210
Charles 283
Cicero 105
Claubry, H. F. Gaultier de 272
Colin, J. 272
Cotton, F. A. 64
Couper, A. S. 39
Courtois, B. 74, 292

Dali, S. 214,
Dante 78
Davy, E. 286
Davy, H. 32, 274
DeBruijn, H. 196
Demokrit 223
Deville, H. S.-C. 54, 289
Döbereiner, J. W. 267, 311
Dürrenmatt, F. 284

Earnshaw, A. 95, 252, 320
Eco, U. 33
Eggert, J. 271
Ehrlich, P. 76
Einstein, A. 199
Empedocles 300
Epikus 223
Archduke Johann 70

Fischer, E. 292
Fogden, M. 265
Fogden, P. 265
Folin 145
Freud, S. 303

Ga'far al-Sadiq   277
Gahn, J. G.   65
Gans   173
Gay-Lussac, J. L.   73
Gilbert, L. W.   272
Glansdorff, P.   251
Goethe, J. W. von   23, 25, 29, 60,
   110, 123, 126, 131, 210, 216, 218,
   232, 260, 264, 282, 300, 327
Göttling, J. F. A.   236
Gogh, V. van   66
Greenwood, N. N.   94, 252, 320
Guldberg, C. M.   121

Händel, G. F.   45
Halpap, H.   186
Hammurabi   27
Harington, C. R.   73
Hauck, H. E.   186
Heeren, F.   92
Hermbstaedt, S. F.   96
Hesse, H.   90, 94, 289
Heumann, K. G.   318
Hoffmann, E. T. A.   45
Hoffmann, R.   5
Hofmann, L.   41,
Holmes, E. L.   144
Huxley, T. H.   212

Jeremiah   27
Isaiah   37, 205
Jorgensen, C. K.   145
Joule, J. P.   237

Kant, I.   50, 92, 306
Karmach, K.   92
Kautsky, H.   170
Kekulé, F. A.   39
Kepler, J.   240
Khalid   18
Klaproth, H.   96
Klee, P.   217

Knebel, K. L. von   223
Koch, R.   76
Körner, T.   321
Krätz, O.   177, 179, 320
Kunckel, J.   219

Landolt, H.   270, 274
Lauth, T.   217
Lavoisier, A. L.   83, 84, 203, 236, 266
Leibnitz, G. W.   42, 188, 203
Lessing, G. E.   5
Lewis, G. N.   120
Lichtenberg, G. C.   15, 88, 139,
   196, 266, 279, 325
Liebig, J.   13, 39, 54, 254, 311
Titus Lucretius Carus   223, 308

Mandelbrodt, B. B.   7
Mahler, G.   183
Maleachi   234
Manet, E.   64
Mann, T.   7, 23, 86
Marks, T. J.   75, 272
Meier-Leibnitz, H.   324
Mendelejew, D. I.   225
Merck, H. E.   39
Meyer, V.   243
Michelangelo Buonarotti   304
Montaigne, M. de   262
Morgenstern, C.   80
Müller, C.   294
Musil, R.   114

Napoleon III.   31, 57
Nernst, W.   180
Neufeldt, S.   144
Nicolai, F.   96
Nietzsche, F.   245, 251, 256, 270

Oppenheimer, R.   80
Osborn, M.   256
Ostwald, W.   IX
Ovid   102

Paracelsus   60, 294
Pedersen, C. J.   104
Peitgen, H.-O.   7
Pemberton, J.   277
Peter von Lothringen   258
Petron   254
Picasso, P.   230, 249
Plancy, C. de   315
Pliny the Elder   318
Poggendorf, J. C. P.   311
Porta, G. della   188
Priestley, J.   82, 236, 266
Prigogine, I.   257
Proust, J. L.   61

Reichardt, C.   119
Reni, G.   203
Rimbaud, J. N. A.   62
Rio, A. M. del   62
Robert   282
Roscoe, H. E.   279
Rouelle, G. F.   135
Ruby, S. L.   75, 272
Runge, F. F.   1

Sagan, C.   273
Scheele, C. W.   83, 236, 311
Schiller, F.   286, 325
Schönberg, A.   155
Seftström, N. G.   62
Seligmann, K.   258
Seneca, L. A. d. Ä.   84
Shakashiri, B. Z.   IX, 125, 193, 195
Sinowjew, A.   240
Skrabal, A.   271
Spedding, F. H.   144

Stahl, G. E.   83, 282
Stöckhardt, J. A.   10, 15
Stromeyer, F.   272
Synesius   18

Teitelbaum, R. C.   75, 272
Thénard, L. J.   38
Thilorier, M.   132
Thomas Aquin   IX
Thompson, S. G.   144
Thomson, W.   237
Trithemius   275
Turner, W.   256
Tutenkhamen   275
Twain, M.   86
Tyndall, J.   223

Valery, P.   65
Verne, J.   31
Vinci, L. da   108, 214

Waage, P.   120
Walden, P.   237
Whitman, W.   68, 142, 221
Wiegleb, J. C.   10, 96
Wilkinson, G.   63
Willstätter, R.   128, 185
Wimmer, F.   296
Winkler, C. A.   191, 225
Wöhler, F.   20, 31, 54, 84, 254, 296
Woodward, R. B.   184

Yogi Berra   54

Zhabotinsky, A. M.   256

Subject Index

acetaldehyde   311
acetic acid   64, 118, 126, 306
– ethyl ester   306
– pentyl ester   309
acetone   73, 79, 94, 120, 186, 315
acetylene   286
acid base reaction   126
activity   120
adsorption184
alcohol   34, 106, 129
-t-butanol   206
– ethanol   20, 68, 106, 118, 124,
   230, 270, 292, 301
– methanol   106, 118, 177, 260, 300
– 1-pentanol   181
– 2-pentanol   308
aldehydes   311
aluminum   31, 54, 137, 232,
– alum   137, 321,
– $Al(H_2O)_6^{3+}$ 138
– $Al(H_2O)_6Cl_3$ 233
– Al_2I_6 32
– $Al(OH)_3$ 138, 234
– $Al(OH)_4^-$ 138, 234
aluminum foil   234
– fluorescent   212
aluminum powder   297
aluminum sheet   232, 295
aluminum shot   58
amber   102
4-amino-p-benzoic acid   212
ammonia   80, 98, 106, 127, 189
ammonia solution   80, 98, 106, 292
– concentrated   80, 120, 292
– diluted   94, 98, 311

ammonia vapor   80, 106
ammonium
– NH_4Cl 29, 51
– $(NH_4)_2Cr_2O_7$ 80, 102
– NH_4NO_3 29, 245
– NH_4SCN 95, 112, 245
– $NH_4H_2PO_4$ 3
– NH_4VO_3 63
amphoteric behavior   137
amylose-iodine complex   74, 100,
   272
aniline   52
anion-exchange resin see resin
anode   8, 229
anthocyanes   128
anthranilic acid   309
– anthranilic acid ester   310
antimony   60
– $SbCl_3$ 60
– SbI_3 60
aromatic principles   308, 313
atomizer   95, 143
auramine hydrochloride   52

barium
– $Ba(NO_3)_2$ 29, 46
barking dog   254
battery   228
– with citrons   228
– with potatoes   228
Belousow-Zhabotinsky reaction   258
bencoic acid   118
– bencoic acid ester   118
benzaldehyde   311
benzoyl peroxide   52

betain dyes   181
bioluminescence   208
bis(2,4,6-trichlorophenyl)oxalate,
  TCPO   206
bis(2,4-dinitrophenyl)oxalate,
  DNPO   206
bismuth
– $Bi(NO_3)_3$ 307
bleaching   92
blood   82
borax   304
boric acid   304
bromine   73
Brønsted acid   136
buffer mixture   127
2-butanone   108

calcium
– CaC_2 287
– $Ca(CH_3COO)_2$ 303
– $CaCl_2$ 244
– $CaCO_3$ 184
– CaO 243
– $Ca(OH)_2$ 243
capacitor   228
carbon dioxide   129, 251
carotine   184
castor oil   178, 203
cathode   8, 228
cell voltage   28, 229
cerium
– $Ce(NH_4)_2(NO_3)_6$ 258
charcoal   46, 52, 298
charge transfer complex   75, 207
chemical garden   23
chemoluminescence   188
– indicator
– 9,10-bis(phenylethinyl)anthracene,
  BPEA   206
– 13,13'-dibenzoanthroyl   194
– 9,10-diphenylanthracene, DPA
  206

– violanthrone   199
– rhodamine B   206
– rhodamine6G   194
– rubrene   194
chemotherapy   76
chloride ion   121
chlorine   73, 197, 280
chlorine-hydrogen reaction   279
chlorophyll a and b   184
chromatogram   184
chromatography
– paper   3, 142, 184
– thin-layer   186
– column   183
chromium
– $Cr(II)$ 114
– Cr_2O_3 79
– chrome alum   323
cigar   239
cobalt   108
– $Co(II)$ 114
– $CoCl_2$ 99, 120
– $CoCl_4{}^{2-}$ blue   101, 121
– $Co(H_2O)_5{}^{2+}$ 101, 109
– $Co(SCN)_4{}^{2-}$ 109
– Co-Sm   240
– $K_2Co(II)(SiO_3)_2$ 108
cocktail   176
coins   294
– minting of   294
column for chromatography   184
complex equilibrium   120
copper   15, 27, 98
– $Cu(II)$ 114
– $Cu(CH_3COO)_2$ 16
– $CuCl$ 249
– $CuCl_2$ 120
– $[CuCl_4]^{2-}$ 121
– $[Cu(H_2O)_4]^{2+}$ 121
– $Cu_2[HgI_4]$ 250
– $[Cu(NH_3)_4]^{2+}$ 122
– Cu_2O 67

– CuSO$_4$ x 5 H$_2$O   67, 97, 177, 321
– Cu(II) tartrate   67
copper mirror   15
copper ore   15
corrosion   25
crown ether complexes   103
crown ethers   102
crystal formation   323
crystals   321
cyclohexene   73

desmolase   85
– katalase   85
3,6-diaminophenthiazine   217
dichloro methane   118, 194
diethyl ether   60, 73, 184, 293
dimethyl formamide   118
dimethyl phthalate   206
dimethyl sulfoxide   118
dinitrogen monoxide   255
disproportionation   38, 87
distilled water   94, 177, 264
distribution coefficient   182
distribution law   180
Döbereiner cigarette lighter   266
dropping pipette   221
dry ice   131
EDTA   68
electrode   228
electrons   224, 229, 235
electrophilic   118
endothermic   245
enthalpy   35, 245
– dissociation   32
– free   245
– reaction   243
entropy   245
ester   307
– synthesis   309
exothermic   30, 248
exothermic reaction   26

felt-tip pen   186
ferroin solution   258
fire   45, 300
– bengalic fire   48
– dancing   300
– green   45, 48
– red   45, 48
– yellow   48
fireworks   29, 45
five-color game   112
flame   30, 40
– green   30
– purple red   39
– violet   32
fluorescence
– blue   218
– dyes   206
– golden yellow   220
fomaldehyde   68, 192, 311
food colorings   3
formic acid   118
fountain   139
fractals   7
fruit ether   308
fuchsinesulfuric acid   312

gallium   225
gas generator   291
gel, burning   303
germanium   225
glacial acetic acid   70
glucose   76, 128, 262
gold   18
green house effect   251
gum arabic   101
gummy bear   35
gunpowder   297

H$_3$O$^+$-ion   129
hemoglobin   208
heptane   74, 114
hexaaqua complexes   115

hexane   74, 114
hoarfrost   232, 316
hydrazine   14
hydrochloric acid   63, 93, 268
– concentrated   120
– diluted   93, 124, 192
– medium concentrated   63
hydrogen   266, 280
hydrogen carbonate   129
hydrogen peroxide   38, 64, 67, 83,
   118, 189, 264
hydrolase   85
hydroquinone   85, 110
hydroxybenzenes   110
– 1,2-di-hydroxybenzene   110
– 1,3-di-hydroxybenzene   110
– 1,4-di-hydroxybenzene   110
– 1,2,3-tri-hydroxybenzene   110
– 1,3,5-tri-hydroxybenzene   111
2-hydroxybenzoic acid   111
hydroxylammonium chloride   50
hypochlorite   197

I_5^- -anion   74
icing sugar   48, 184, 208
ignition powder   297
indigo   52
inks, sympathetic   97
iodic acid   277
iodine   31, 60, 73, 178, 277, 292
iodine-starch complex see amylose-
   iodine complex
ion exchange column   144
iron   25, 224
– lattice structure   27
– alum   323
– citric acid complex   214
– Fe(II)   114
– Fe(III)-O bridged complex   111
– $FeCl_3$ 94, 110
– $K_3[Fe(CN)_6]$ 210
– $K_4[Fe(CN)_6]$ 3, 95, 113, 211

– FeO   26
– Fe_2O_3 24, 58, 297
– $Fe(phen)_2(CN)_2$ 118, 258
– $Fe(SCN)_3$ 95
– $Fe(SCN)(H_2O)_5^{2+}$ 95, 113
– $FeSO_4$ 216, 258
– block   295
– column   27
– oxalic acid complex   24, 211
– oxide   58
– passivation   27
– powder   297
– self igniting   25
isoamyl valerate   309

Jahn-Teller effect   115
Joule-Thomson effect   237

lactose   52
Law of Constant Proportions   61
Law of Mass Action   120, 307
lead   12
– $Pb(CH_3COO)_2$ 5, 10
– PbI_2 5
– $Pb(NO_3)_2$ 11
– Pb_3O_4 238
lead tree   10
lewis acid   118
– acceptor number   118
ligand exchange   105
ligand field   107, 115
light petroleum   183, 284
light scattering effect, see Tyndall
   effect
liquefaction of gases, see Joule-
   Thomson effect
liquid air   226
litmus solution   126
luciferin   208
luminol   189, 208
luminophore   219
magnesium   32

– $MgBr_2$ 219
– $MgCl_2$ x 6 H_2O 164
– MgI_2 32
magnesium powder   297
magnesium ribbon   58
malic acid   256
membrane   24
manganese   64
– Mn_2O_7 33
– $KMnO_4$ 34, 39, 64, 88, 103, 224
– K_2MnO_4 103
– Mn(II)   114
– MnO_2 34, 35, 86
– $MnSO_4$ 264
melting point apparatus   311, 319
mercury   18, 223
– $Hg(CH_3)_2$ 18
– $HgCl_2$ 232
– HgI_2 238, 250
– $Hg(NO_3)_2$ 249
– HgS   18
– $Hg(SCN)_2$ 20
mercury mirror   18
methane   301
methylene blue   76
micro pipette   186
mineral water   37, 130
mist see smoke
Mitscherlich test   201
multi phase system   177

negative electrode see cathode
neutrino   225
nickel
– Ni(II)   114
– $NiCl_2$ 106
nickel complexes   105
– $[Ni(NH_3)_6]^{2+}$ 106
– $[Ni(H_2O)_6]^{2+}$ 106
– $[Ni(H_2O)_m(NH_3)_n]^{2+}$ 106
nitric acid
– concentrated   27

p-nitroaniline red   52
nitrogen triiodide   292
nitrogen, liquid   236, 301
p-nitrophenol   68
NO_2-N_2O_4-equilibrium   238

oscillation   257
overhead projector   7, 134
oxalic acid dichloride   194
oxalic ester   205
oxidating agents   16, 35, 53, 189
oxide layer   28, 31, 233
oxygen   80, 280
oxyhydrogen gas   289
– in soap bubbles   289
ozone   18
ozone generator   19

paper   94, 142
para red   52
paraffin   178
– oil   178
paraldehyde   311
pasteur pipette   3
patterns, formation of   133
petrolether   284
pH-indicators   124
– list of   124
pharaoh's snake   20
phenolphthaleine   12, 124, 303
philosophers stone   18
phosphorus   42, 188
– P_4O_{10} 43, 202, 308
– red   43
– white   42, 188, 295
photodissociation   281
picture development   210
– solutions for   210
– stencil   210
pictures   1
piezoelectric spark generator   284
piperidine   319

plastic freezer bag   212
platinum rod   261
platinum spiral   80
platinum sponge   267
polystyrene foam   306, 315
potassium
− $KBrO_3$ 256
− $KClO_3$ 35, 46, 294
− K_2CO_3 135, 177, 192
− $K_2Cr_2O_7$ 102, 224
− $K_2[HgI_4]$ 250
− KI   6, 73, 249
− KIO_3 70, 270
− $KMnO_4$ 34, 39, 64, 88, 102, 224
− K_2MnO_4 102
− KNO_3 52, 298
− $KNaC_4H_4O_6$ 67, 311
− KOH   50
− KSCN   108, 120
pottash, see K_2CO_3
n-propanol   33
protolysis   136
prussian blue   95, 113, 211

quicklime   243
quinone   85

radical chain reaction   271
rainbow colors   125
red prussiate of pottash   210, 322
redox indicator   77, 259
− indigocarmine   262
redox pair   85
redox potential   74
redox reaction   30
redox system   77
reduction   14
− electrolytic   7, 12
− photochemical   216
resin   144
− anion exchange   144
reversible process   307

round bottomed flask with
   condenser   60
salicylic acid   308
− methyl ester   309
− sodium salt of   216
salt   3, 100
salts
− acidic   135
− basic   135
seignette salt   311
semicarazide   311
semicarbazone   313
silver   7, 13
− $Ag_2[HgI_4]$ 250
− $[Ag(NH_3)_2]^+$ 14, 312
− $AgNO_3$ 8, 14, 120, 249, 311
− Ag_2O 14
− Ag_3N 9
silver cutlery   234
silver mirror   13
silver tree   10
singlet oxygen   196
smoke   50
− blue   53
− green   53
− red   53
− white   53
soap bubbles   289
soap solution   290
soda, see Na_2CO_3
sodium
− CH_3COONa 164
− $NaAsO_2$ 70
− $NaBO_3$ 208
− NaCl see salt
− Na_2CO_3 112, 164, 301
− Na_2CrO_4 102
− $NaHSO_3$ 68
− $NaIO_3$ 238
− $NaNO_3$ 120
− Na_2O_2 37
− NaOCl   44, 197, 203

- NaOH   65, 76, 118, 179, 263, 306
- Na$_3$PO$_4$ 208
- Na$_2$SO$_3$ 68, 260
- Na$_2$S$_2$O$_3$ 70, 228
- Na$_2$SO$_4$ 184
sodium hydroxide solution   90
- diluted   90, 124
solvents   117
- polar-aprotic   118
- protic   118
sparkler   297
special magnets (Co-Sm)   240
sprinkler, see atomizer
starch   70, 264
stars and stripes   143
stationary state   259
strontium
- Sr(NO$_3$)   46
structure
- polymeric   43
- self-organized   133
sublimation   317
suction flask   60, 183, 311
sudan III dye   176
sulfur   18, 46, 123, 239, 247, 298, 318
- CS$_2$ 43, 73, 256, 295, 319
- modifications   320
- monoclinic   320
- orhtorhombic   320
- prismatic   320
- polycatena sulfur   320
- SO$_2$ 46, 299
sulfuric acid   34, 63, 70, 90, 224
- concentrated   33, 47, 63, 70, 114, 118, 224, 257, 304
- diluted   90, 224, 268, 291
superconductor   240
- pellet   240

thermite mixture   59
thermite procedure   54
thermocromism   248
thymolphthalein solution   68
thyroxine   74
tin
- SnCl$_2$ 219
toluene   60, 73, 183, 319
trichloromethane   102, 118
triethyl amine   181
3,4,5-trihydroxybenzoic acid   111
triplet oxygen   196
two-color chemoluminescence   191
Tyndall effect   222

UV-light   212

vanadium   62
- V(II)   114
- ions   63
viscosity   320
volume contraction   120

waterglass   23
whitener   218
wood wool   37

xantophylles   185
m-xylene   177
p-xylene   177, 319

zink   10, 29, 63, 247
- granulated   267
- Zn(II)   114
- ZnO   30, 248
- ZnS   248
zink dust   29